THE VOYAGE OF THE *WATER WITCH*

THE VOYAGE OF THE

WATER WITCH

A SCIENTIFIC EXPEDITION TO PARAGUAY AND THE LA PLATA REGION
(1853 - 1856)

ROBERT D. WOOD, S.M.

LABYRINTHOS *1985*

The map and illustrations are reproduced from Thomas J. Page's popular edition of the expedition, *La Plata, the Argentine Confederation, and Paraguay* (New York: Harper & Brothers, 1859). The original legends for each are also used.

Portions of this work appeared as "The Remarkable Voyage of the *Water Witch*: The Parana River Expedition of 1853-56," by Robert D. Wood, in *Americas* (April 1966), 4-9.

The cover design is adapted from the April 1966 issue of *Americas.*

Library of Congress Card Catalog No. 85-50095 / ISBN 0-911437-15-0

Printed in the United States of America

LABYRINTHOS, 6355 GREEN VALLEY CIRCLE, No. 213, CULVER CITY, CA 90230

PREFACE

In 1856 Lieutenant Commander Thomas Jefferson Page, of whom you will learn more in the following pages, wrote in his report to the Secretary of the Navy, "No portion of South America has so largely occupied the attention of European statesmen within the last twenty years, as the states of La Plata. . . . These countries are worthy of our highest consideration."

Strangely enough, in spite of the proximity and common interests that should bind us to Latin America, this "highest consideration" has been very slow in coming. Only in recent times, with the "Good Neighbor Policy," the Pan American Union, and the "Alliance for Progress," have North Americans begun to pay some serious attention to these countries, and we are discovering—as Commander Page also did—that "the historical associations of this vicinity are full of interest." Many fascinating events have been overlooked or ignored by historians, and the expedition of the *Water Witch* from 1853 to 1856 is one of them. The following work will, I hope, give it some of the attention it deserves.

Although the narrative often reads like a fanciful adventure tale, it is a true story. The setting is South America's La Plata basin: dramatically involved are Argentina, Uruguay, and Paraguay—especially the last. The geographical names are given as they are spelled today; accents, almost totally absent in the original documents, have been restored to conform to modern usage. In a few instances the original spelling has been retained for local color. Most of the technical and historical details have been relegated to footnotes.

In regard to the many rivers and their left and right banks, the geographical convention is "face downstream": that is, looking toward the river's mouth, the banks are right and left. Thus, in *ascending* the Paraná River the cities of Diamante, Paraná, and Asunción are on the left bank, though they would be on the traveler's right, since the Paraná runs south. As to the countries in which the many cities and towns visited are located, the Index will be of some help.

This work was drawn from a number of archival documents, as indicated in the Essay on Sources. Their exact location is indicated in the References for the historian or scholar.

My appreciation and thanks go to Mr. Elmer Huber and the staff of the Military and Naval Branch of the National Archives, Washington, D.C., for their kindness and cooperation during my research there.

Now let your imagination take you back a century to the days of tall masts and billowing sails—and churning side-wheels, and come aboard the *Water Witch* as she skims the seas and plows the rivers of the La Plata region.

R.D.W.

February 1985

CONTENTS

The Water Witch.

1. OVERTURE TO AN ADVENTURE

Almost every adventure worthy of the name is the precipitate of a combination of an almost impossible dream and a series of fortuitous circumstances. It is likewise true that any interesting enterprise is the result of an individual's belief in himself and his mission. Henry James said, "Ideas are, in truth, forces; infinite, too, is the power of personality; a union of the two always makes history."[1]

No Horatio Alger enthusiast could ask for more exciting elements than those which entered into the adventure of the *Water Witch*, nor any romanticist find a more ideal setting than Paraguay a hundred years ago—remote, mysterious, fascinating in its semi-isolation and seclusion. The dream was to open this remote region to the ships of the United States and to make of the Paraná River a highway of world commerce. This dream was made a reality through the forceful personalities of two very similar men who seized upon the circumstances of the fall of one dictator and the rise of another, and the anxiety of a Confederation and an infant nation to take their places in the world community. And so it happened that the *Water Witch* not only glided into some of the most dramatic moments of South American history, but also made some unforgettable history of her own. Vision and fortune made the outcome inevitable.

Perhaps, after the death of the great isolationist Francia in 1840,* many men had the idea of beating a path to the interior of South America and opening Paraguay's door. One of the first to

*José Gaspar Rodríguez Francia (1761? -1840), known as Dr. Francia or "El Supremo," was born in Asunción. A shrewd legalist, he became active in Paraguayan independence in 1811 and was designated one of two consuls to govern the country. In 1814 he became dictator and three years later assumed the post for life. Almost literally the only educated person in the primitive colony, he met no opposition (having emasculated the small native aristocracy) and compulsively kept himself—and Paraguay—utterly secluded and unenlightened. (Washburn.)

express this idea was the United States representative in Montevideo, R. M. Hamilton, who in 1843 wrote a fourteen-page report entitled "Notes on the Political, Moral, and Commercial, condition of Paraguay: and the advantages offered by that country to Mercantile undertaking, and methods of carrying it, into successful operation."[2] But any ideas that Hamilton or others may have had about opening doors were quickly crushed in the following year when the dictator-president of Buenos Aires, Juan Manuel de Rosas,* closed the Paraná to navigation, stretched a chain across the river, and built batteries of guns along the banks.

Carlos Antonio López,† who had been elected on March 14 of that year to a ten-year presidency of Paraguay, considered the action of Rosas the start of an undeclared war. Perhaps it would be more accurate to say that he felt himself involved in a war that had been going on for some time. Three years previously Rosas had already sided with the rebel Uruguayan ex-president Manuel Oribe,‡ whose troops still held Montevideo in a state of siege. Rosas hoped through Oribe's victory to bring Uruguay, formerly the Banda Oriental Province, back into the Argentine Confederation. But this itself was falling apart.

The provinces of Entre Ríos, Santa Fé, and Corrientes were slowly but surely moving away from the dictator's control. To complicate the situation even more, France and England were openly siding with the anti-Rosas provinces, particularly Corrientes. It was for this reason that Rosas had closed the Confederation rivers to all foreign ships. When Paraguay protested against this strangle hold on her own main artery of access to the outside world Rosas accused the country of aiding Corrientes against him, and even considered the possibility of extending the Confederation to the old viceroyalty boundaries which included Paraguay.

To complete this already complex picture, Brazil succeeded in 1845 in reconquering the revolutionary province of Rio Grande do Sul, and now studied the possibility of advancing once again into Uruguay to deal with Oribe's forces which had been crossing into Brazil's southern province on cattle raids. In addition, Brazil also objected to the closing of the La Plata basin because through it lay the easiest route to her interior provinces, particularly Mato Grosso. Brazil had hoped for some time to make Paraguay an ally against Rosas, but while López actually did assist Corrientes, he was wary of Brazil's real intentions, and until after the defeat of Rosas maintained rather frigid feelings toward the Empire.[3]

Into this thorny and politically complex problem stepped one of the most unusual representatives the United States has ever sent abroad. Edward Augustus Hopkins,§ the son of an Episcopalian bishop in Vermont, was twenty-two years old when he decided to leave the U.S. Navy and seek adventure and fortune on his own.[4] While on duty in Rio de Janeiro he had heard of the hidden land of Paraguay and his romantic and visionary impulses had been stirred to discover more about it.[5] He read all he could about the country and decided it would be his

*Juan Manuel de Rosas (1793-1877) was born in Buenos Aires, of which city-province he became governor, 1829-31, 1835-52. He gained power over other Argentine provinces and became a dictator to match Francia. He waged war against Uruguay (1842-51), closing the La Plata estuary to all trade, but was nevertheless defeated by Brazil and her allies. Rosas fled to England in 1852 to live in exile.

†Carlos Antonio López (1790-1862) had been a consul three years when he became president of Paraguay, 1844-62. Sometimes called "López I," he reversed Francia's isolationist policy, thus making it possible for the *Water Witch* to visit.

‡Manuel Oribe (1796? -1857), born in Montevideo, as a boy joined the Army of the Rio de La Plata. He was one of "thirty-three immortals" who liberated Uruguay. (Living in Buenos Aires, the "thirty-three" raised an army and marched to Banda Oriental—the "East Bank," between the Uruguay River and the Atlantic Ocean, then belonging to Brazil. After three years of war, 1825-28, it became the free state of Uruguay.) Oribe became president (1835-38) but was deposed and fled. With Rosas, General Oribe began a civil war (1842-51) culminating in a siege of Montevideo which failed. He was defeated by General Urquiza.

§Edward A. Hopkins (1822-91) was born in Pittsburgh and educated (without receiving a degree) at his father's school in Burlington. Sailing as a midshipman 1840-45 he was struck by opportunities in Latin America, especially Paraguay. Although he resigned from the Navy (June 1845) he had been court-martialed three times for unruly conduct and was close to dismissal anyway. (Peterson, 247; Ynsfran, 249.)

special purpose and privilege to open Paraguay to the United States. He had in fact many qualities which could have made him extremely successful. He was highly intelligent, enthusiastic and energetic;[6] the advantages of being a good horseman and a fine musician who could sing and play the guitar made him socially popular.[7] But he also had a quick and almost uncontrollable temper, coupled with a pride that expressed itself in an extreme sense of personal honor. His enthusiasm, too, was often only a shade away from impetuosity. Most of all he was in no sense a diplomat—as future events sadly showed.

With his customary energy and directness Hopkins gathered his information on Paraguay and wrote to the Secretary of State, James Buchanan, asking him to persuade President James Polk to name him special agent to Paraguay.[8] On June 10, 1845, Buchanan wrote to Hopkins that the President had decided to send an agent to Paraguay. "The industry and zeal which you have manifested in collecting information on the subject and presenting it before the President, have mainly caused your selection for this mission." The assignment had several purposes. Hopkins was to inquire about the character of the people, the population, the militia, the financial system, the persons in government, and the possibility of trade. He was also instructed to investigate the claims of Buenos Aires on Paraguay and the right to exclusive navigation of the La Plata. He could also tell the President of Paraguay that if a representative would come to Congress the following December the country would probably get a recognition of its independence.[9]

Delighted with his commission, Hopkins sailed from New York on the *Catheline*, arriving in Rio de Janeiro after forty-one days on July 30.[10] He saw at once the entanglement of Brazil in the La Plata affairs, though he thought at first that Rosas was a champion of republican principles.[11] The blockade of Buenos Aires and the fighting in the Banda Oriental made it impossible for him to sail farther, nor would the Brazilian government, not knowing his purposes, give him permission to travel over land. Using his ingenuity, Hopkins posed as a scientific investigator and was given a letter of introduction to the Duque de Caxias,* who was the president of Rio Grande do Sul.[12] When Hopkins arrived there the Duque was away on military matters, so there was nothing he could do but wait. During his prolonged stay he wrote a lengthy account of the political situation.[13]

The crux of the matter, as he saw it, was simply a question of mutual renunciation: Brazil would renounce any claims to Paraguay if Rosas would renounce his claims to Uruguay.[14] It never seems to have occurred to Hopkins that such a "renunciation" would ultimately mean only an exchange of the claims. Before he became too involved in political affairs, however, he was able to leave for Paraguay, posing this time as a writer interested in historical information.[15]

After one month and ten days on horseback, Hopkins entered Asunción in the middle of the morning of November 8.[16] He was received royally, and within a short time he had gained the complete confidence of President López, mainly by telling him that the United States would soon recognize Paraguay's independence.†[17] He then set about to investigate the things indicated in the instructions from Buchanan. As time went on, he came to the conclusion that "no

*Luiz Alves de Lima e Silva, Duque de Caxias (1803-80), was born in Rio de Janeiro. In the 1851-52 war against Argentina he had been commander-in-chief of the Brazilian army. As a marshal he would command Brazilian forces against Paraguay (1866-69). Caxias was prime minister of Brazil 1850, 1856-57, 1861-62, 1875-78. He was the only duke knighted by Pedro II. His father, Francisco de Lima e Silva (1785-1853), had also been a general and statesman, ruling Brazil 1831-35. (Kolinski.)

†This of course was an eventual certainty, but Hopkins at that moment had no authorization to say so. More damaging, however, was his March 1846 letter to Argentina's dictator, where with what Peterson called "appalling impudence" he presumed to tell Rosas he must change his policies: ". . . would it be more agreable **[sic]** to your last moments, when the rapid review of your life passes before you, to be able to say, I have desolated my Country; impoverished it; ruined it, and gloated over the blood of my enemies; than to feel . . . that you die calm and contented, for your Country is happy and at peace, and you alone had caused it?" Secretary of State Buchanan had to disavow the letter. Hopkins was virtually ignored in Washington the following six years. (Peterson, 249-50, 252.)

Carlos Antonio López.

Government of the Civilized world knows anything about Paraguay, my own when I left the United States, perhaps least of all."[18]

Hopkins learned that Paraguay was divided into eighty-five counties and that there were nineteen Indian settlements. Crime was almost unknown; there had been only five executions in four years, and there were only fifty people in jail. This was a great change from the days of Francia, who "made it the standing rule of his life, to shoot every man of more than ordinary influence, abilities or information."[19] Everything pleased the enthusiastic agent, and he concluded that even though there were "many traces of the federal system of old Spain remaining amongst the Paraguayans,"[20] the country was "the most united, the richest, and the strongest nation of the new world" next to the United States.[21] The people of Paraguay, in fact, were familiar with the famous men of the United States, whereas in the interior of Brazil, Hopkins said, the people didn't even know where North America was![22]

As to commerce, the possibilities were superb, and there was "every prospect that an American Company will secure the steamboat navigation in the waters of Paraguay, under the Paraguayan flag."[23] The foundation for this belief was the fact that "the President is most anxious to see American steamers in the waters of Paraguay, as soon as the difficulties are arranged."[24] Nor was Hopkins blind to the possibilities of amassing a personal fortune by taking advantage of the relatively undeveloped rich resources of the country. And thus began the strange dual role that Hopkins was to play for many years and which was eventually to be his undoing. He tried to be commercial agent and diplomat at the same time and failed in both.

President López' interest had been aroused nevertheless, and when Hopkins asked for a fifteen-year monopoly of steam navigation on the rivers of Paraguay for a company he would form, he promised to think about it.*[25] For Hopkins this was as good as assent, and he established a small office in Asunción in front of which he flew the star-spangled banner in Paraguay for the first time on Christmas day, 1845.[26]

The great problem remaining for Hopkins was the question of Buenos Aires, which he had been told to investigate, and in which he now had a personal interest. He wrote to President Buchanan asking to be made Commissioner to Paraguay with powers to act as mediator between Buenos Aires and Paraguay.[27] Without waiting for an answer Hopkins suggested the idea to López, who was favorable and agreed to withdraw his troops from Corrientes if Rosas would promise not to attack Paraguay.[28] Hopkins set off at once for Rio de Janeiro where he hoped to find passage to Buenos Aires. He succeeded—and it was undoubtedly a puzzled American who heard him say he had come to settle the problems of Paraguay when he told William Brent, Jr., the United States representative in Buenos Aires. Brent, however, spoke to Felipe Arana, the Minister of Foreign Affairs for Buenos Aires, who accepted the proposal.[29] The outcome of the talks was what might have been expected.

Rosas was willing to recognize the independence of Paraguay as a province on the same footing with other provinces, but demanded its reincorporation into the Confederation.[30] Hopkins knew that such terms were absolutely impossible, so he himself went to Rosas to present Paraguay's propositions personally. He was kindly received—and adamantly refused. Disheartened, Hopkins wrote a long letter to López praising Paraguay, blaming Buenos Aires, and reluctantly admitting that the mediation was a failure.[31] He informed the Paraguayan president of events taking place in Buenos Aires, and advised him against accepting the services of General Fructuoso Rivera,† who was reported ready to offer them to Paraguay.[32]

*Hopkins originally proposed to form a company capitalized at $300,000. He found a group of Rhode Island financiers, headed by Samuel G. Arnold, lieutenant-governor of the state, who had himself visited South America and saw its potential. Hopkins also wanted to organize a manufacturing company in Paraguay and an agricultural school for Paraguayan boys; these did not materialize. (Peterson, 254-55.)

†José Fructuoso Rivera (1790? -1854), born in Uruguay, was one of the "thirty-three immortals" who freed the country in

While Hopkins was contemplating his next move a communication from Secretary of State Buchanan burst the great bubble of his pipe-dream. He was informed in forceful and firm language that he had gone far beyond the limits of his instructions in promising Paraguay recognition of independence, in assuming a diplomatic character, and in attempting an unauthorized mediation. He was ordered to return to the United States at once.[33] From Rio Hopkins sent a letter of farewell and friendship to López asking, oddly enough, that he correct the mistake a Paraguayan newspaper had made in reporting that Texas had recently been forced to join the United States.[34] To Buchanan he sent a long report on La Plata political affairs, defending Paraguay and requesting its recognition by the United States government. He concluded his letter by saying that he had traveled 18,000 miles, 2400 of them on the backs of 300 horses![35]

Hopkins was not a man to give up easily. In the next few years he returned to Paraguay five times. He explored the Paraguay and Paraná rivers, and got to know the country thoroughly. During this time he continually pushed for formal Paraguay-U.S. relations. So evident was his interest that in 1851 López named him special minister of Paraguay to the United States! At the same time he offered a bribe and a suggestion: if Hopkins would get United States recognition of Paraguay, he could have the monopoly he had asked for. Why not go to the States and return with a steamer? Paraguay would welcome some sort of expedition—say, for scientific or nautical purposes.[36] To Hopkins the idea was excellent. There was only one drawback: the chain across the Paraná.

The goddess of fortune sometimes smiles at the most unexpected moments, proving thereby that fact is often stranger than fiction. No news could have been more welcome throughout the La Plata region than the announcement that General Justo Urquiza,* for years the strong support of Rosas, had turned actively against him. By June the nine-year siege of Montevideo had been raised, and in October Oribe capitulated. All attention turned to the tyrant in Buenos Aires. In December 1851 Urquiza crossed the Paraná into Entre Ríos with 28,189 men. Six weeks later the forces of Rosas were decisively defeated at Monte Caseros.†[37] The governors of the Argentine provinces met at San Nicolás and appointed Urquiza Provisional Director of the Argentine Confederation. After the stormy years of the Rosas regime the most welcome rainbow was the decree of Urquiza, dated August 28, 1852, opening once more to the ships of the world the waters of the La Plata and its tributaries.[38]

The downfall of Rosas brought immediate effects in the United States. On April 28, 1852, the Secretary of State, Daniel Webster, sent instructions to John S. Pendleton, the U.S. Chargé d'Affaires in Buenos Aires, and Robert C. Schenck, U.S. Minister in Rio, to try to conclude treaties of friendship and commerce with the Argentine Confederation, Uruguay, and Paraguay.[39] Hopkins, who admittedly knew more about Paraguay than anybody else in the United States, was appointed Consul to that country. His persistence had paid off. Nor had he forgotten his last interview with López.

On the night of January 13, 1852, Hopkins in New York had spoken before the American Geographical and Statistical Society, where he had stressed the value and importance of the United States being the first nation to send a steamer—he recommended a side-wheeler—up the

1825. After leading a revolt he became first president (1830-35), a post he held again in 1838-42. Ousted from the Uruguayan presidency by the Oribe-Rosas combination, Rivera had been defeated by Urquiza the previous year (1845) at the battle of India Muerta. He fled to Brazil.

*Justo José Urquiza (1800-70) had been governor of Argentina's northern Entre Ríos province, 1842-54. As a general he defeated Rivera (1845), but when Rosas closed the La Plata, isolating the northern provinces, Urquiza deserted him. He defeated Rosas (1852) to become the country's commander-in-chief of all forces and first Argentine constitutional president, 1854-60.

†After crossing the river with the assistance of Brazilian troops, Urquiza's army spent six weeks marching to Corrientes province (Argentina), where Monte Caseros is located.

Paraná River into Paraguay.*[40] His words had the desired effect, for on May 11 S. DeWitt Bloodgood, secretary of the Geographical Society, sent a memorandum to the Secretary of the Navy encouraging and backing the project.[41] Meanwhile, Hopkins was busy organizing The United States and Paraguay Navigation Company.[42]

By fortunate coincidence the memorandum sent to the Navy Department struck a responsive chord.† John P. Kennedy,‡ "one of the ablest naval secretaries,"[43] had already conceived the idea of sending out several survey expeditions, particularly to the Pacific areas around China and Japan.[44] Why not include an expedition to the La Plata region also?

The privilege of sailing on this expedition, which was to be equally as historic as the one which went to Japan, fell to the lot of a remarkable navy lieutenant who had actually asked for something else. Thomas Jefferson Page§ was a Virginian filled with the spirit of enterprise and adventure. He was a good mariner who could command and who was brave and calm in the face of danger. In addition, he had considerable scientific knowledge and literary culture, and his personal writings show a certain poetic sensibility.[45] When he heard of the proposed Pacific expeditions he applied for command of one of them.

On September 24, 1852, Secretary Kennedy sent Page two letters. In the first he explained that while he was aware of Page's "meritorious standing in the Navy . . . such a command justly belongs to an Officer of higher grade and longer service than yours," and that Commander Ringgold had been chosen for the China Sea expedition. However, he added, "I think you

*Hopkins as a midshipman had learned much about the superiority of steam power on rivers, where tides and contrary winds had always made them difficult for sailing vessels. He also wanted to dazzle Paraguayans with its new technology. England and the United States both want to sell manufactured goods to Paraguay, he warned the Geographical Society: England uses steam, America only sail. Finally, ". . . the appearance of an American river-steamboat in these [La Plata] waters would increase our exportations to these regions a million of dollars the first year and . . . would double every six months thereafter. . . ." he claimed. (Hopkins, "Memorial," 33, 40.)

†America through the 1840's and '50's (the administrations of Presidents Zachary Taylor and Millard Fillmore) was eager to open new—foreign—markets for its manufactured goods. Furthermore, the Monroe Doctrine (promulgated 1823), prohibiting intervention by European powers in North or South America, and "Manifest Destiny," the belief that expansion across North America—if not more—was America's democratic duty toward lesser endowed societies, set up a perfect climate for exploratory investigation.

The first U.S. Exploring Expedition to the Antarctic (1838-42) had taken years to organize and staff—and finance by a Congress unsure of its implications and advantages to the public. Charles Wilkes (1798-1877) was its forceful if excitable commander, driving its six ships to the Antarctic (where Wilkes Land is named after him), Tahiti, the Samoas, Australia and New Zealand, Tonga, Fiji and the Sandwich Islands, and the northwest coast of America. The expedition became a strong precedent for other voyages. In the Civil War Wilkes again received notoriety when he abducted two Confederate agents (James Mason and John Slidell) from the British ship *Trent*; because of international law the U.S. government was obliged to release them.

(Stanton; Ponko, 19-32; Harrison, 175.)

‡John P. Kennedy (1795-1870), a novelist and member of the House of Representatives, strongly believed that exploration and scientific endeavor had legitimate places in government. He was Secretary of the Navy less than a year (July 1852-March 1853), but during that time he dispatched three expeditions: Ringgold's (North Pacific), Lynch's (West Africa), and Page's. He decided that, unlike the others, the La Plata exploration required no appropriation from Congress, as he had full authority to send naval vessels into Argentine waters. He charged the expedition's cost—$220,000 (not counting its two ships)—against regular naval allocations. (Ponko, 15, 109; Harrison, 193-95.)

§Thomas Jefferson Page (1808-99) was born in Virginia, the eighth son of Mann and Elizabeth (Nelson) Page, both descendants of distinguished Virginia governors. Page was appointed midshipman 1827 and served in the West Indies and in Mediterranean and Brazilian waters. On coastal survey work (1833-42) under Ferdinand R. Hassler (1770-1843), a Swiss engineer who had taught mathematics at West Point, he demonstrated a taste for the scientific curiosity and methodological investigation then in vogue. "Exploration appealed to Lieutenant Page," Ponko noted, simply.

In 1848-51 Page commanded a brig in the Navy's Far East command. "Here," said a biographer, ". . . he realized the need for a surveying expedition in the China seas, for the benefit of commerce and whalers, and upon his return proposed it to the department." Such proposals were common as part of any application for a novel or unique assignment: as opportunities in the peacetime Navy were few, the scientific expedition was a viable alternative for promotion. (*Dictionary of American Biography*, xiii 140; Harrison, 195.)

entitled to the second Command.* . . . There will be two vessels of inferior size to the *Vincennes*, either of which will be offered to your command as you may select."[46]

The first letter, as Page later wrote, was flattering—but the second letter interested him more. In it the Secretary asked him if he might be willing to command an expedition to the La Plata region of South America.† He accepted the second offer at once, and some days later received a long communique from Kennedy which informed him that "the President has directed a small steamer to be dispatched to the La Plata for the purpose of exploration and survey of the upper streams above their falls, and to which service you have been assigned in command of the United States steamer 'Water Witch.' "[47]

The purpose of the exploration was to determine the practicability of navigation and to discover the courses, extension, and fish production of the rivers.‡ Examination was to be made not only of the shores but also of the interior of the adjacent country to get exact information on the nature and extent of agriculture and the probability of commerce. Astronomical, weather, and magnetic observations were to be made, and specimens of zoology, botany, mineralogy, and geology—in short, natural history—were to be collected wherever possible. Scenes and events worth recording were to be preserved by "Daguerrean" [sic] and camera impressions.§[48] The instructions stressed the fact that although the *Water Witch* was a man-of-war, the purpose of the expedition was to extend the bounds of science; the ship's guns were only for defense in case of extreme emergency, and she was to avoid all interference in any foreign affairs. The Secretary hoped that from this expedition many wonderful things would result.

The *Water Witch* had been built originally at the Washington Navy Yard in 1843 as a small steam tug for harbor use. Two years later it was taken to Philadelphia and rebuilt, and saw service in the Gulf of Mexico during the Mexican War. It soon became a battered-up boat, but the machinery was still in good condition, so once again it was taken to the Washington yard and rebuilt,‖ this time as a side-wheel steamer man-of-war, "the only vessel of that kind we ever had in the navy," said a historian.[49] The ship was especially constructed to experiment with the new Morgan paddle wheel¶ which would be thoroughly tested by a long voyage on the high seas.[50]

*That is, to be second-in-command to Ringgold. Cadwallader Ringgold (1802-67) became a midshipman 1819 and was promoted to commander 1849. After several voyages around the world he participated in the Wilkes expedition and later (1849-50) surveyed the California coast.

The North Pacific Exploring and Surveying Expedition (1852-56), which Page declined, reached China March 1854. After protecting foreigners there against rioting, Ringgold contracted intermittent fever (malaria) which left him temporarily insane; Page might thus have become expedition commander—as happened to John Rodgers, who accepted the post.

Having himself sailed around the world—he had just returned from China—Page was not making an idealistic or romantic decision. He obviously wanted a command position of any size. (Page, xix; Ponko, 109.)

†Lieutenant William F. Lynch had requested the La Plata expedition, but without success. He had already explored the Dead Sea and Jordan River (1847-49), his reports extremely popular in America. Lynch instead was instructed to explore the west coast of Africa (1852-53), in preparation for the possibility of returning freed slaves there. (Ponko, 33-60, 199-205, 244.)

‡In his instructions to Page Kennedy said the expedition's principal objective was commercial, i.e. ascertaining river conditions. But as the Latin American countries were often suspicious of mercantile motives, the Paraguayan government was to understand it to be a scientific trip. Congress, however, generally had not approved voyages of a purely scientific nature; a commercial application had to be included. (Harrison, 196, 200-201.)

§The Daguerrotype photographic process (using a light-sensitive emulsion on glass plates) dates to about 1839. The camera or camera oscura was a portable "dark room" using its lens to focus an image on a surface where an artist could trace it on paper. It was a popular device for scientific expeditions.

‖The *Water Witch* was 180 feet long with a 23 foot beam. The ship's engine had come from an earlier vessel with that name, condemned in 1851. The eight-year-old engine, designed for a ship half the size, was going to be heavily taxed in its new task. (Ponko, 110, 273.)

¶The "Morgan Excentric" paddle wheels were 19 feet in diameter, 5 feet wide, and dipped 42 inches under the water's surface. They weighed 24 tons. The eccentric component was an additional "feathering" mechanism which tilted each paddle or "box" vertically as it entered the water, increasing its efficiency. The wheels were from the Morgan Iron Works of New York.

Charles Morgan (1795-1878), starting as ship's chandler, had a fleet of steamers along the Atlantic, around the Gulf of

Since Hopkins had suggested a side-wheeler for the river exploration, the *Water Witch* would serve a double purpose. Made of wood, the ship was of 378 tons burthen with a nine-foot draft—an important factor where river travel was concerned. She was equipped with three bronze howitzers.[51] For her trip to La Plata she would have to be specially prepared, and this was to be Page's first concern.

To outfit a vessel for a voyage of long duration, knowing that much of the material and supplies will not be available abroad, and having to consider in addition the specialized demands of a scientific expedition, is no easy task. Page was fortunate in having the assistance of Lieutenant Matthew Maury,* head of the Naval Observatory and himself a scientist, who helped select the instruments for the trip.[52] The foresight with which Page prepared for his cruise is indicated in the answers of the Secretary of the Navy to his petitions for various purchases. On November 30 Page was authorized to buy guns, ammunition, and small equipment such as scissors, gauze, axes, acids, specimen bottles, geological hammers, three pounds of insect pins, and three pounds of fishing hooks.[53] Two weeks later he was given permission to purchase two double-barrel guns with Maynard's primers and India-rubber covers, and four leather demijohn bottles, and to go to New York "for the purpose of examining certain Boats in the Navy Yard, to ascertain whether they are suitable for the duty on which the *Water Witch* is destined."[54]

Page's preparations extended beyond instruments and equipment. For the purpose of advancing knowledge and immediate reference, the *Water Witch* had her own library. Some of its books included the recently released work of Sir Woodbine Parish, *Buenos Ayres and the Countries of La Plata*, Robertson's *Letters on Paraguay* and *Francia—Reign of Terror, Natural History of Paraguay, Elementary Course of Geology, Mineralogy and Physical Geography* (1850), *Elements of Chemistry*, Gray's *Botanical Text Book*, Griffith's *Medical Dictionary*, Stephenson's *Marine Surveying*, Biddell's *Magnetical Instruction*, d'Orbigny's *Historique du voyage dans l'Amerique Meridionale*, and various atlases. In all there were forty-eight books.[55]

On January 12, 1853, Secretary Kennedy wrote to Page: "The Steamer 'Water Witch' having been reported ready to leave the Navy Yard (Washington) you will at the earliest day practicable, proceed in her to Baltimore for the purpose of receiving on board her crew, and a coal engine; after which you will proceed to the Navy Yard Norfolk for your stores."[56]

For a voyage such as the *Water Witch* was to undertake, the crew was a matter of great importance. Page had previously been told that some regular seamen would be transferred to the ship,[57] but these were the days when an officer often had to find his own ordinary seamen, landsmen, and boys. It was in these groups that the turnover was most frequent, with some

Mexico, and in the "California trade," serving the Pacific via Panama and Nicaragua. In 1836 he acquired the foundry and machine shop and turned out, besides marine engines and boilers, a great number of paddle wheels and walking-beam mechanisms. See also n., p. 78.

(Page, 26, 30-31; Ponko, 109, 273.)

*Matthew F. Maury (1806-73) was one of the most distinguished American scientisits of his time. He entered the U.S. Navy in 1825 and soon criticized its training methods; somewhat in response, the Naval Academy at Annapolis was formed in 1845. But having made a number of enemies (and lamed by a stagecoach accident) Maury was relegated to the Depot of Charts and Instruments and the Naval Observatory, both of which he headed, 1842-61. (They were later merged.)

Lieutenant Page had served at the Naval Observatory 1844-48; he and Maury were both Virginians, both of scientific disposition. For the expedition Maury drafted comprehensive instructions for Page, but—when Secretary Kennedy furnished his own elaborate orders—they were put aside and sent later. See n., p. 11. Correspondence shows that Maury and Bloodgood of the Geographical Society had already been discussing the La Plata when Hopkins spoke. The society was composed largely of merchants and shipping magnates, whose interest in foreign places was more than academic.

Not long before, Maury had similarly assisted another scientific expedition—one he actually conceived: Lieutenants William Herndon (Maury's cousin) and Lardner Gibbon were sent (1851-53) to explore the headwaters of the Amazon. Their work led ultimately to Brazil's opening of the river to international trade. See also n., p. 86.

(Ponko, 61-64, 244.)

sailors "jumping ship" at the first port of call. Page picked his crew, writing later, "Some had been confided to me as 'hard cases' by tearful mothers . . . and a 'man-of-war' is considered an asylum for all such."[58] However, for the majority he found their conduct "highly creditable and all that could be desired."[59]

Page was more than pleased when passed midshipmen William L. Powell, William H. Murdaugh, and Edmund W. Henry wrote to him asking him to take them on the trip and support their applications to the Secretary of the Navy.* He gladly did so, since these men were skilled in some of the sciences with which the expedition was concerned.[60] In addition to them, there were two other officers of equal rank with himself:† Lieutenants Edward Donaldson and Daniel Ammen. Ammen was not anxious at all about his assignment to the trip and joined the ship only two days before she sailed. He left the expedition after one year, but it is to his accounts that we owe much interesting information. Dr. Robert Carter was assigned as surgeon to the ship, and George P. Welsh was appointed Acting Master. The all-important engineers were Second Assistant Engineers William Lamdin,‡ R. C. Potts, and T.B.C. Stump. On February 5, 1853, Lieutenant Page sent the list of the crew to the Secretary of the Navy for approval.[61]

Page had brought the *Water Witch* from Baltimore to Norfolk in nineteen hours, and wrote to Kennedy, "I was much pleased to see her 'stand up' so well."[62] Now all that remained was to make the last-minute arrangements and preparations. Page had thought of the possibility of building a small steamer in Asunción for the exploration of small rivers, creeks, and so forth. He arranged with Secretary Kennedy to have the small boats and boilers sent to the La Plata region by freighter[63] and to make requisitions to pursers of other vessels in the vicinity of the *Water Witch* whenever necessary. At the same time he received permission to negotiate "for small sums" on the bankers, Baring Brothers & Company of London,§ who had representatives in several South American cities where the *Water Witch* would visit.[64]

Only a few days before departure Page was surprised to receive a letter from Edward Everett, the Secretary of State. It informed him of Webster's commission to Pendleton and Schenck of the previous year, a commission which had not as yet been carried out. Page was given "full power" to arrange a treaty with Paraguay himself, using the U.S.-Costa Rica treaty of 1851 as a basis. If Pendleton and Schenck wanted to go with him to Asunción he was to take them, but it was not necessary to do so for the sake of the treaty. Free navigation of the Paraguay River was the main aim. This mission he was to keep secret until it was accomplished.[65]

Everything was ready. On the morning of February 8, 1853, Page went aboard with his second son, John, twelve years old, who had been given permission to go along on the expedition.[66] The last stores—919 pounds of powder and 140 tons of coal—were stowed away.[67] At 1:50 p.m. the ship weighed anchor; at 4:00 the harbor pilot left her. Under fair skies, with pleasant north winds billowing her sails, the *Water Witch* wedged through the swells of the open sea.

*The ship's passed midshipmen (cadets who had completed their training and were awaiting commissions as officers) petitioned that they be appointed acting lieutenants, which the Secretary of the Navy could do in non-naval assignments. But having treated the La Plata expedition as routine (see n., p. 7), Kennedy had to decline. Page had endorsed the petition, and later gave them "field" promotions. "Landsmen" (above) were sailors on their first voyage.

†Although their ranks were the same, Lieutenant Page as commander of the vessel was its captain (which also designates a higher rank), and was occasionally so addressed.

‡Ponko spelled Engineer Lamdin as "Lamden." (Ponko, 110; Page, 595.)

§International English bankers and fiscal agents for the United States.

Asunción.

2. SIDE-WHEELS SOUTHWARD

Steaming steadily southward, the *Water Witch* watched other sails slowly slipping from sight. On a smooth sea she made good time, passing Puerto Rico on the seventeenth and coming to her first anchorage in the harbor of St. Thomas island a little after noon the next day. She remained here four and a half days while her engines and boilers were cleaned, supplies of fresh beef and vegetables brought on board, and five hundred gallons of water shipped.[1] During this first stopover Captain Page wrote enthusiastically to Kennedy:[2]

> The performance of the Water Witch—her paddles and boilers—being an Experiment with us, and involving a degree of skepticism, in the minds of ~~many,~~ at home amounting to a certainty of their failure—met my most sanguine expectations—she has been tried—as a "sea boat"—and not found wanting; and her paddles and boilers have proved successful.

He wrote the same thing to Maury, with whom he kept in touch during his whole expedition.*

When the ship had refueled—she used ten tons of coal a day[3]—she pointed her slender prow southward again and in succeeding days saw Guadeloupe, Dominique, Santa Lucia, and the Barbados Islands drop below the horizon. On March 4, with the temperature now in the mid-eighties, the *Water Witch* churned into Demerara (Georgetown), British Guiana. She was the first American man-of-war ever to enter the port.[4] Some members of the crew were assigned to

*Maury's interest in the expedition was far from one of scientific detachment. In connection with his interest in the Amazon (see n., p. 86) he wanted Page to discover whether the La Plata headwaters were connected with the Amazon; he wanted all winds and currents recorded—even noting on which side of rivers driftwood accumulated. He also instructed Page to send home every Paraguayan article of commercial value—seeds, roots, plants, fruits, by-products, etc. (Harrison, 196.)

fixing the fore sheet, which had fallen overboard the day before and been cut by the wheel. Some of the officers met a Dr. Turner, a brother of the English chemist, who took them about twenty miles inland toward Berbice, where they saw a herd of Burmese cattle and fields of yellow cotton. Ammen reflected, "It was altogether an interesting place, and would have been a pleasant one but for the prevalence of yellow fever."[5]

There was a Baring Brothers establishment in the port city and Page took advantage of the opportunity to make his first withdrawal for funds—$2000—"to meet bills."[6] On March 13 the *Water Witch* left Demerara and four days later the pilot from Le Père station brought the boat into Cayenne, French Guiana.[7] The ship's stay was brief, but there was little to be seen anyway, except for the impressive double row of royal palms leading up to the government offices and several companies of French soldiers keeping guard over thousands of exiles languishing away in this land of living death. One prisoner seized a small boat and followed the *Water Witch* as she left, pleading to be taken on board.[8] He was soon lost in the distance as the ship unfurled her topsail and sped toward the equator.

When the boat had hit the equatorial stream she encountered an extremely strong contrary current. While in Cayenne Page had met the captain of a French man-of-war who let him copy his own charts, enabling him to avoid the worst of the stream.[9] By doing so the ship was able to reach Maranham (the English equivalent of Maranhão, later named San Luis) on March 24, five days after leaving Cayenne, in spite of continual rainy weather. Because of the constant pull against the current it was necessary to overhaul the engines and raise the wheel shafts,* and to slush the masts because of the frequent rain. These tasks, in addition to taking on fuel and food, took two weeks. At the end of the first week the men were sent on liberty—and typical sailor reactions occurred. George Blizard and William Brown were confined in double irons for drunkenness and insolence, and Samuel Chaney didn't return until the next day. But return he did, little knowing that one day he would be part of the subject of an international conflict. On April 7, 1853, the *Water Witch* set sail once more.[10]

Clouds and flashes of lightning greeted the ship as she steamed into Pernambuco (Recife) on April 13. When the storm broke the men found the city "a charming-looking place from the sea," protected by a natural mole† of smooth coral rock.[11] While the crew was engaged in painting the ship and taking on supplies some of the officers went to the country residence of the American consul.[12] Before leaving, Page drew five hundred pounds from the Baring Brothers firm.[13] The men were pleasantly surprised to discover that the pilot who took them from the harbor spoke English.[14]

Eight days later, on April 23, Rio de Janeiro's Fort Santa Cruz and the cathedral came into sight. The previous day's run of 196 miles remained the record day throughout the whole voyage.§[15] Ordinarily the ship averaged 144 miles daily. When she reached Rio after forty days and twenty-one hours, she had logged 5913 miles from Norfolk.[16]

Rio proved to be a port of disappointments and difficulties. Two days after arrival there Page wrote to Maury, "I find that I am robbed of one of the brightest jewels of the Expedition by Pendleton."[17] Pendleton and Schenck, together with Sir Charles Hotham of Great Britain and

*Ammen perhaps had some mechanical savvy that Page lacked. The heavy weight of the paddle wheels had caused their shaft bearings to shift. Jacks had been furnished for raising the wheels, but they proved inadequate. Ammen's solution was to bring two harbor barges alongside the ship, swamp them and lash them to the wheels; when bailed out and refloated they raised the wheels a half inch—enough to replace the bearings. Page does not credit Ammen with this idea. Slushing (below) means to apply a protective coating like linseed oil. (Ammen, 248.)

†Pier or breakwater.

§No mileage in the three logs of the voyage supersedes this figure. The "cathedral" may instead have been the old Jesuit church on Castelo Hill.

the ministers of France and Sardinia, had succeeded in concluding a treaty of "friendship, commerce and navigation" with Paraguay.[18] Francisco Solano López* had been appointed to act for Paraguay.[19] Pendleton was informed that the treaty had been ratified in Asunción on March 12, and he in turn replied that Paraguay had been given a twenty-one gun salute by the U.S. flagship *Jamestown* in Buenos Aires.[20]

This was only the first disappointment. The second came when Commander John Pope of the *Vandalia*, stationed in Rio, demanded that Page give him one of the new chronometers brought especially for the expedition. Page could not argue—nor did he care to—with a senior officer,† but he wrote at once in protest to the Secretary of the Navy, James C. Dobbin, who had succeeded Kennedy in office on March 7.[21]

On April 26 Page wrote a letter to Ferdinand Coxe, Secretary of the Legation in Rio, acting in the absence of Schenck. He explained the purpose of the expedition and asked the diplomat to obtain the cooperation and permission of Brazil for the exploration of the interior rivers, particularly the Paraguay. Coxe immediately relayed the message to the Brazilian Minister and Secretary of State for Foreign Affairs, Paulino José Soares de Souza, adding that the expedition would be of the highest significance to the commercial and scientific worlds, and that Brazil would benefit greatly from it. The Secretary was informed that the *Water Witch* would leave on April 30, but should an answer be impossible before that time it could be reached in "Monte Video" [sic] or Buenos Aires.[22] Page's third disappointment was that the answer did not arrive before the departure date.

On top of these troubles, virulent yellow fever, which had been brought to Rio from Africa in 1849, broke out on board.[23] Fortunately it was kept under control, and most of the crew could be employed in loading the ship with such interesting items as mustard, pepper, beeswax, jack-knives, black silk handkerchiefs, molasses, and—taking advantage of Brazil's best—two barrels of coffee.[24] Page, meanwhile, added to his list of annoyances by buying a chart and afterwards discovering that the whole coast of South America was fifteen miles too far east.[25] He took solace, however, in drawing 260 pounds from Baring Brothers.[26]

Skies were clear on April 30 when the *Water Witch* hove anchor into place and started on the last long leg of the journey south. The sea was heavy, and strong currents fought the steamer's steady progress. In spite of this she was still making about 164 miles a day at an average of 6.4 knots.[27] Then, three days out from Rio, the cause of the continuous heavy sea became clear. The ship ran head-on into a "pampero," a violent southwest wind that turned the ocean into a mass of foaming mountains. The jib-boom broke, and men worked through an inky night of rushing winds and splashing spume to get it back on board.[28] "The waves broke over the *Water Witch* like a cataract," said Page.[29] He wrote to Maury, "A more *dangerous* sea I have never experienced; especially for a low-sided vessel as is the Water Witch."[30] He added that the storm had been a real test for the side wheels and had proven the Morgan action superior.

When the sea became relatively calm again the sailors saw what looked in the distance like large sand bars. They proved to be huge schools of small shrimp, disturbed by the storm.[31] The remainder of the voyage was uneventful, and on May 7, 1853, the ship moved into the outer

*Francisco Solano López, "The Marshal" (1827-70), was López' son and would succeed him as president, 1862-70. At age 17 he was already a brigadier general in his father's army. Part Guaraní (like his father), he was proud of his Indian heritage, made rousing speeches to his troops in the native language, and like Francia and his father had a misanthropic streak that became all consuming. (Washburn, i 389, 405.)

†Technically, the *Water Witch* was part of the Brazil squadron, equipped for its exploratory tasks with just a few extra books and instruments. Having (in Page's words) "no scientific corps on board," the mission would lack any great urgency or cachet—certainly in the eyes of the squadron commander. John Pope (1798-1876) commanded the *Vandalia* 1853-56. (Page, 62-63.)

harbor of Montevideo. No sooner had she entered the "inner roads" than the captain of the port came on board and put her under an eight-day quarantine. So far as the crew was concerned the time was not wasted. While the ship stood in the outer harbor the Spanish flag at her foremast and the American at her main, the masts and gun carriages were cleaned, the engines put in order, and supplies taken aboard. Most of the equipment received was for the building of a small steamer: adzes, riveting hammers, copper nails, wedges, caulking mallets, soldering irons, and pitch.[32]

Shortly after the quarantine ended Commodore Isaac McKeever,* commanding the U.S. ships in Montevideo, came on board the *Water Witch* to offer whatever help he could to facilitate the expedition.[33] Page undoubtedly appreciated the gesture and shortly after was wishing that Brazil had been equally as amiable. He received Minister Soares de Souza's reply to Coxe's letter—and it was not exactly what he had expected. The Minister informed him that the river would be open to Page to the port of Albuquerque, and that the president of Mato Grosso and other imperial agents would give him their cooperation. He could not go beyond the point mentioned, however, because that portion of the river had not as yet been opened to foreign nations, and there was no complete agreement as to the right of ownership of the territory. An expedition into the waters might establish a precedent "which might be prejudicial to the Empire."†[34]

This frustration to his plan of going up the Paraguay River as far as possible irritated and somewhat angered Page. Something of a solution came by coincidence when he went to the War Department in Montevideo to obtain a map of Uruguay. The officer in charge was Colonel Wenceslao Paunero, a brother-in-law of José Ballivián,‡ the president of Bolivia. On hearing of Page's problem, Paunero produced a map showing that the territory above Albuquerque belonged to Bolivia, even though it had been claimed by Brazil. Any expedition into Bolivia would be more than welcome.[35]

Page became convinced that Bolivia owned the right bank of the river and that Brazil had no right to refuse him ascension beyond Albuquerque. To Maury he wrote,[36]

> Brazil is grasping, and *worse* than the Chinese in her exclusive policy. She has stepped over into Bolivia and claimed a strip of country along the Paraguay from Albuquerque up to the Monument[?], to which—it is very clear—she has not the shadow of right. . . .

He added that Brazil had better "abandon her 'Dog in the Manger' policy—or she will be forced to it,"§ and concluded by asking Maury to get a map of Bolivia from England because "it is on *our* side."[37] It was an English-made map that Paunero had shown him.

On May 24 the *Water Witch* left Montevideo at 2:30 p.m. with several passengers, some of whom were sailors being transferred from the *Congress* to the *Jamestown*, and arrived in Buenos Aires at 8:00 the next morning. She "hoisted the Buenos Ayrian flag," an action which, had not

*Isaac McKeever (1793-1856) had been a captain since 1838 and was an old friend of science. In 1839 he visited Wilkes' ship at Valparaiso (Chile), lending him sone men and equipment, and persuading Wilkes to take his nephew, Lieutenant Edwin J. De Haven, on the exploring trip. (Stanton, 113.)

†López would express the same objection; see n., p. 54.

‡José Ballivián (1804-52) as a general fought under Andrés Santa Cruz; he had been president of Bolivia 1841-48. His son Adolfo (1831-74) became president 1873-74.

§John P. Harrison, a Latin American specialist at the National Archives, judged Page "an explosive force" to match Hopkins, citing this letter as an example of dispatches and letters which "bristle with indignation whenever his wishes were not immediately granted." "Page fairly yearned for glory," he added. Like Hopkins too, Page wanted to "open" Paraguay to trade, using, if necessary, the forceful tactics employed with China and Japan. Any "non-industrialized" nation, said Harrison, would find this expression of Manifest Destiny threatening. Brazil, however, proved to be no problem. (Harrison, 198-99.)

the American flag also been in evidence, might have brought about a battle.[38] Page learned that Buenos Aires was being besieged and blockaded by General Urquiza, who was trying to force the city to come into the Confederation and to adopt the new Constitution.[39] The *Water Witch* had run both the outer and inner blockade squadrons.

The next day, accompanied by Pendleton, Page went into Buenos Aires where he found "few or no structures of great architectural merit . . . except the churches and convents," though the streets were paved with granite and there was an abundance of flowers "of which these people are extravagantly fond."[40] The two men rode about a mile and a half into the country to San José de Flores, where Urquiza had his headquarters in a small and simple house. The General was asleep when they arrived, but the visitors were considered of sufficient importance to awaken him. A short time later he appeared before them, "a stout, well-formed person, of medium height, with fine, piercing eyes, and frank countenance,"[41] "a person of polished courteous manners, affable and pleasing in conversation."[42] Page told him of the purpose of the expedition and of his intention to ascend the Paraná to Paraguay, and to explore, at the same time, wherever possible, the rivers of the Argentine Confederation. Urquiza understood at once and seemed pleased with the prospects of such an expedition.* He immediately ordered his secretary to issue instructions to the authorities of the different provinces to do whatever they could to facilitate the work of the *Water Witch.*[43]

Page was pleased with the outcome of the interview and felt the way was clear except for the barrier presented by Brazil, but he wrote with a note of finality to Maury, "Her claims will not exclude *me*—that is a *fixed fact.*"[44] At the same time, writing to Secretary Dobbin, he recorded a momentous occasion in Argentine history:[45]

> On the same day, on which I was presented to him [Urquiza], he received at the hands of Commissioners—the Constitution which has been fraimed [sic] by the Congress of the Confederation now assembled at Santa Fe. It has been submitted to the City for adoption, as it also has been to each province; and it seems to be well understood that the provinces—including that of Buenos Ayres—will unanimously adopt it; but it is as generally believed that the *City* of Buenos Ayres will reject it. . . . Should the City reject this Constitution she will deserve the punishment that may be visited upon her.

During these decisive days the *Water Witch* was making her final preparations for the trip to Asunción. She took on iron work for the small steamer and made a short run up the Uruguay River to find wood for fuel.[46] Anxious as he was to start for Paraguay, Page wanted to wait for the *St. Andrews* coming from Philadelphia with coal for the *Water Witch.*[47] The decision to delay a bit led the *Water Witch* to play an important role in the history then being made.

A dramatic turn of events was taking place. Supply ships began moving into Buenos Aires, and the blockaders deliberately fired so as to miss these "running storehouses"; nor were the ships intercepted at any point. Urquiza hoped to starve the city into submission, but with supplies coming in this would be impossible.[48] Then came the climax: it was June 20, a memorable day for several reasons. Early in the morning the men on the *Water Witch* had dressed the ship for the anniversary of Queen Victoria's accession to the throne of England. At 2:30 that afternoon they saw one of the blockade steamers come in and anchor with the inside squadron. At 8:04 both squadrons weighed anchor and left for the inner roads. The blockade was lifted.[49]

What had happened? Page had partly prophesied when, in writing to Maury, he said there were "villains who are filling their pockets on this state of things. . . ."[50] Admiral Coe had sold

*Although he had defeated Rosas (February 1852) Urquiza could not persuade Buenos Aires province—the richest of all, because of its maritime commerce—to join his confederation. He hoped that his new free river navigation policy—aided perhaps by *Water Witch* discoveries—would make the other provinces less dependent on Buenos Aires. (Harrison, 192-93; Ponko, 108.)

out the Confederation squadron to the city of Buenos Aires for $480,000, of which he kept $208,000 for himself. He then took refuge on the only "neutral" vessel nearby—the *Jamestown.*[51] Shortly after, Page received orders from the captain of that ship telling him not to leave without permission and to be ready to tow the flagship over "the flats" on July 2. Page obeyed reluctantly out of respect for rank, and on the appointed day the *Jamestown* left for Rio, supposedly for caulking. Admiral Coe was still on board.*[52]

The situation in Buenos Aires had grown tense. Urquiza was expected to attack the city at any time. Page wanted to leave, not only to avoid any involvement in local affairs, but also because he thought the Paraná would be at the low water season before he could get up the river.[53] His plans were frustrated once more by an urgent note from Pendleton on July 3 asking him to remain in port for the protection and possible evacuation of American citizens, since his was the only U.S. vessel in the vicinity. Schenck supported Pendleton's petition.[54] It was true that "a marked consideration for the lives and property of foreigners characterized the course of Urquiza throughout this siege," said Page,[55] but anything could happen. So Page waited. The following day the harbor resounded with the booming of guns fired from the fort on shore and from the French, Brazilian, and Spanish flagships. It was a tribute to the *Water Witch* for the Fourth of July![56]

In his note Pendleton had written, "Events of importance, and of a decisive character in respect to the condition of this City . . . are . . . at hand."[57] A week later Pendleton informed Page that the diplomatic representatives of England, France, and the United States had succeeded in bringing about peace by convincing Urquiza to lift the siege of Buenos Aires.[58] "It is part of the present plan, that the foreign men of War in port may convey the Provisional Director and his escort to the neighboring port of Gualeguaychú."[59] If the siege were to be lifted, Urquiza and his troops had to be assured a safe escape. This task fell to the *Water Witch.*

On July 13 the ship moved two miles north of Buenos Aires to Palermo, where Urquiza and his staff consisting of one general, six colonels, seven majors, seven captains, eight lieutenants, fifteen civilians, and thirty soldiers as guards, along with "a quantity of baggage," were received on board.[60] Page gave his private cabin to the wife and daughter of the general accompanying Urquiza who, gentleman and soldier that he was, took berth in the ward room.[61] Two English steamers, the *Locust* and the *Trident*, carried the remaining troops.[62] Five days later the passengers disembarked at Point Basilio, Page and Captain Day of the *Locust* personally seeing them safely ashore. Urquiza expressed his eternal gratitude for the services of the *Water Witch*, and was later to say on several occasions that the ship had saved his life.†[63]

Her historic mission accomplished, the *Water Witch* started back for Buenos Aires, but even on the way she was again to be of service. Both of the British steamers went aground, and it took three days to transfer their crews ashore and pull the boats back into the water.[64] Nor was this the end of the whole affair. No sooner had the ship reached Buenos Aires on the evening of the 22nd than Page received a note from Pendleton and Schenck asking him to take them to Gualeguaychú so they could conclude with Urquiza the treaty of friendship and commerce which

*John (Juan) H. Coe (1806-64), whose American parents had settled in South America, at age 17 became a sailor and participated in the revolutions of Chile and Peru. He then went to Buenos Aires and helped liberate Uruguay; as an admiral he fought with Rivera in Montevideo and then returned to Rosas, who made him squadron commander of the Argentine Confederation. He then fled to the United States, as Page suggests here. The *Jamestown* commander was Captain Samuel W. Downing, senior officer of the Brazil squadron.

†Urquiza's guard totaled 85 soldiers. The entire party consumed so much food Page wrote to Secretary Dobbin for extra funds (denied). This rescue and Page's subsequent visits to Urquiza nevertheless cemented a long friendship later very advantageous to Page. See n., p. 86. (Ponko, 112.)

they had been commissioned to execute, and which circumstances in Buenos Aires had made impossible.[65]

Page did not feel he could refuse an official request, so once again the *Water Witch* steamed up the Uruguay River to a point just below the mouth of the Gualeguaychû. Urquiza had sent a special carriage to meet them, and the two diplomats persuaded Page and some of the ship's officers to accompany them to Urquiza's *estancia.* It was a brisk mid-winter morning, and the fifty-mile ride across open country was invigorating and refreshing.[66] It was the first time they had traveled any distance on land, and they saw many quail and deer and—for the first time—ostriches* "on their native heath."[67]

The *estancia*† was enormous. Within its bounds were 70,000 sheep, 40,000 cattle, and 2000 horses. The house itself was made of stone and was long and flat, of one story, except for the two octagonal towers at either end of the edifice. It was furnished in the finest European style.[68] The guests were entertained in a manner "most distinguished and hospitable," and when the treaty was signed they were served a sumptuous dinner with "fine wines and cigars."[69]

On August 2 the *Water Witch* returned to Buenos Aires and Page sent a report of his mission to Dobbin, adding with satisfaction, "The treaty which they have just concluded could not have been made, but for the Services, which the Water Witch has rendered on this occasion."[70] The following day the ship took Schenck back to Montevideo where Page found the long-expected *St. Andrews* waiting for him. He also met Lieutenant James A. Moore, who had been assigned to the *Water Witch* and who had recently come from Baltimore, bringing with him, according to a prearranged plan, a twelve-pounder howitzer which was a gift for President López from President Franklin Pierce.[71]

For three weeks the *Water Witch* rode at anchor in the Montevideo harbor while she was completely overhauled and resupplied. Except for routine problems such as "several men in double irons for drunkenness," the days were ordinarily uneventful.[72] But monotony was never long a companion on the expedition. On August 19 Page received a letter from the U.S. consul in Montevideo, R. Hamilton, telling him that Hopkins, the consul-designate for Asunción, had arrived in Montevideo on a ship belonging to Buenos Aires which was flying the American flag. Hamilton objected to this abuse of the nation's emblem and asked Page to do something about it at once.[73] Page immediately sent a note to Hopkins telling him he had no right to fly the American flag on a foreign vessel and to remove it from the mast at once. Hopkins replied that this was all "much ado about nothing," and, after all, he was an American official and saw no reason he couldn't use the country's flag on the ship carrying him. Page's answer was prompt and peremptory: ". . . you will recognize this as an order to haul down the American flag now flying at the peak of the Buenos Ayrean propeller Constitucion."[74]

But Hopkins demurred. It might reflect on his official position if he backed down now. Couldn't he wait until he got back to Buenos Aires? Page, probably much disgusted with the whole affair by this time, gave in to the request, insisting, however, that the flag not be flown in the port itself.[75] To Dobbin, a week later, Page wrote, "I have told Hopkins and the Buenos Ayrean government that if I see the Constitucion with the United States flag I will take possession of her."‡[76]

*The South American ostrich is called the rhea, of which there are two species. Darwin sent some fragments of one back to England and was honored to give it his name, *Rhea darwini.* (Von Hagen, 191-92.)

†"A cattle-farm, but not exclusively so, where there is some cultivation of the soil." (Page, 52n.)

‡The symbolism of the flag was full of emotion for Hopkins too. He later wrote, "having already shown the American flag for the first time in Paraguay (1845), which flag was made there under my direction, and is now in the family of the late Hon. S. G. Arnold of Rhode Island, who hoisted it at Providence as Governor, on the news of the surrender of Gen. Lee; I now took it for the first time to several different ports of the Argentine and Uruguayan Republics." (Ynsfran, 250.)

Just before leaving Montevideo Page purchased the hulk of an old brig to serve as a kind of barge in which to carry a supply of coal.[77] With the hulk in tow, the *Water Witch* returned to Buenos Aires where she remained for five days and where several seamen were discharged at their own request and new ones taken on. Finally, in the afternoon of August 31 the ship steamed out into the La Plata–destination, Asunción.[78]

Passing the guardia Tres Bocas.

3. UP THE PARANA

To a mid-nineteenth century traveler the long journey up the Paraná River might have been a tedious experience, made tolerably pleasant perhaps by the splendid scenery and the promise of novel diversion in some of the settlements along the way. To the men of the *Water Witch*, however, it was much more than a mere excursion.

It was more, even, than an exciting adventure into the heart of a continent which to most of them had been only a name in a book, as surrounded by mystery as the African Congo or the Celestial Kingdom of China. Undoubtedly thousands of eyes had already seen what they were about to see, but they were the eyes of travelers for the most part. Only the early explorers who had looked through the eyes of discovery would have known the real spirit of the expedition. And even here there was a difference.

The *conquistadores* had come to find and to claim and to exploit. The men of the side-wheel steamer, searching with the eyes of science, had come to classify natural history, expand the bounds of science, contribute to the world's knowledge, and make that world a next-door neighbor to an almost unknown region. Everything was an object of interest, and constant and keen observation became part of the daily routine. The officers especially were aware of the fact that every mile was another line on a page of history, because the *Water Witch* was the first ship of its kind to attempt to ascend the river. Yet, not only did the men on board make history, but they also helped preserve it. Their jottings, journals, and letters are pictures of the times in which they wrote, and the reconstruction of the era is every bit as intriguing as the opening of an Egyptian tomb or the discovery of a clarifying document.

The La Plata waters in many places are full of shoals and dangerous sand banks,[1] and it took the *Water Witch* an hour to find the channel.*[2] The island of Martín García which stands at the head of the La Plata like a little Gibraltar was only an obscure form in the rain, and soon the whole landscape disappeared into the night. By morning the ship had reached the delta land of the Paraná,† vast flat country cut up by little streams, with large tracts of rushes, solitary except for the loon and the fish hawk.[3] It took some time to find the main artery into the river. As the ship progressed, the landscape changed from marsh land to sand hills to dense growths of willow and alders.[4]

Progress was slow because of the uncertainty of the river channel and because of the constant soundings which were being made as part of the expedition's purpose. San Pedro, a little town of 2000 and still in the province of Buenos Aires, was the first sign of civilization,[5] though the area around it was full of deer and water turkeys.‡[6] Eleven miles above the town was a place called Obligado, and at this point the river narrowed to a little less than a half mile in width. It was here that Rosas had built a battery of guns and stretched a chain across the river. The French with broadsides alone destroyed the battery, and the English had hoisted and cut the chain.§[7]

As the side wheels pulled onward, the scenery continued in never-ending variety, from marsh lands to forested banks to bluffs which reached a height of eighty feet at San Nicolás,[8] the city of the first post-Rosas "congress." Even the river itself, which was now from three to four miles wide, continually changed its character, the bottom shifting from silt to sand of heavy gray and fine white,[9] while the channel moved from one bank to the other. At times it seemed scarcely like a river at all, but more like the sea surrounding an archipelago, so numerous and continuous were the islands in it. Each of these, in turn, was a "wilderness of foliage and flora," "fringed with aquatic plants and . . . willows whose long, drooping branches dipped gracefully into the waters."[10]

Toward evening on September 6, the ship anchored off Rosario amid thunder showers and very vivid lightning.[11] This city was a life-line to the interior, and there were many ox-cart caravans from Tucumán and Córdoba.[12] An American engineer, Mr. Allen Campbell, was studying the possibilities of building a railroad from Rosario to Córdoba.[13] During the rainstorn the hulk carrying the coal broke loose from the boat and drifted down the river, and some hours were lost in recovering it. There were always some small problems. An incident which occurred the next day showed that the rigid Navy practices of the times were not absent from even an expedition such as this. When a sailor named Hoffman refused to work, Powell triced him up,‖ "which made him quite willing to resume pumping after an hour or two of suffering."[14] At sunset the ship anchored opposite "a very pretty church and the village of San Carlos."[15]

The next day was rainy and cold, and there was little to be seen or done. On the afternoon of the ninth, the wheels of the *Water Witch* came to a stop off Diamante. Historically and scientifically it was an interesting spot. It was here that Urquiza had crossed with his army to defeat Rosas, and the brilliance of the victory had led him to change the name of the place from Punta Gorda to Diamante.[16] Along the shore the scientific searchers found masses of coral in the

*The La Plata River was named by Sebastian Cabot, who in 1526 traded here with the Payaguas, a tribe of Guaranís, and got some silver—Sp. *plata*—jewelry (which had come from Peru); "Payaguas" was corrupted by the Spaniards to Paraguá, Paraguay. Really only an estuary, an arm of the Atlantic Ocean, the La Plata is 180 miles long and 170 miles wide—so wide at both ends its shores cannot be seen by each other. (Washburn, i 42.)

†In Guaraní, "Resembling the Sea" or "Great Water." The Paraná is 500 miles long. It is presently being dammed.

‡This is the blackish-brown snakebird, a good swimmer and diver, subsisting on fish; so called for its snakelike neck.

§Supporting Rivera (n., p. 6), the French and British—maintaining their commercial interests in the La Plata area—had aided in disrupting Rosas' blockade. (Page, 71, 81.)

‖To trice, "hoist," was usually to tie a prisoner overhead by the wrists. The U.S. Navy had abolished flogging in 1850. (Ponko, 13.)

form of boulders with the lower part stratified, and there were abundant specimens of gypsum in the cliffs.[17] It was three days before the ship started off for the capital of the Confederation.

Once the ship had gotten under way Page decided not to stop at Paraná or Santa Fé, but to continue until it became necessary to anchor and cut more wood for fuel. These cities could be visited on the return trip. So the steamer paddled on almost 150 miles until she reached La Paz on September 14, 1853. The country around was rich and beautiful, but uninhabited, and the town itself "a small meanly built place"[18] of 200 people who lived in adobe dwellings.[19] Page called it "a miserable place!" and noted that the only person with any energy at all was a priest who had come nine months before from Madrid, and who had already succeeded in growing a fine vegetable garden, but his example had not as yet aroused any imitators.[20] In the region around La Paz was an abundance of swan, turkey, wild hen, and partridge, and Ammen shot a great quantity of doves "of very large size."[21] The following day he, Dr. Carter, and Mr. Bamberger, a passenger from Buenos Aires to Asunción, went horseback riding and shot some bucks, all of which escaped; they managed to bring back "one very fat deer." On the same day Lieutenant Moore and six men were transferred to the coal hulk which was left in La Paz, and the *Water Witch*, resupplied, got under way.[22]

The men on the ship continued to keep track of the topography, noting the tree-thick bluffs along the banks, and especially the abundance of the valuable kapok tree.* The ever-present islands were filled with wild peach trees in early spring bloom, and the air was redolent with delicate odors from these and other flowering plants. "Poets would have revelled in it as a scene of paradisaical beauty," Page remarked.[23] On the afternoon of the sixteenth Ammen, who had shot two *carpinchas*† and was preparing them as specimens, noted a strange phenomenon:[24]

> . . . about three I saw a well defined dark cloud which I thought arose from a fire; about five I was surprised to see the cloud approach in the form of a vast multitude of grasshoppers which really darkened the air and appeared to extend for miles. Some of them lighted on the decks and were put away as specimens.

A "very good specimen" was gotten the next day when Ammen shot a 268-pound deer swimming in the water. The little library on board came in handy, for they found the animal described by d'Orbigny,‡ who had killed two of them in 1828.[25]

Passing clouds and moderate breezes accompanied the *Water Witch* as she steamed past Goya, where the river narrowed to a mile and a half, the banks still thickly fringed with trees.[26] The cultivated fields which had begun to appear from time to time in areas near settlements seemed neglected and barren. Page reflected, "Civil wars have left their desolating mark on this part of Corrientes."[27] On September 20 a broken wheel strap and the need for more wood determined Page to anchor off Bella Vista.[28] The town of 1000 inhabitants was situated on a bluff overlooking the river. Several officers of the ship met a Mr. Hall, an Englishman who had been there for seven years. He and Mr. Yeatman, an American, and "an Irish Catholic priest who was quite tipsy, and a good many others" were invited on board.[29] The steamer was to be a wonder wherever she went.

The town itself had little to offer ("everyone is miserably poor"), but the surrounding area was full of wildlife, particularly parakeets in huge numbers.[30] Fifteen miles above the city, on

*The seeds of the kapok or silk-cotten tree (also called the ceiba) are covered with silky fibers used to fill mattresses and life preservers and for insulation.

†The capybara, a South American rodent somewhat like the related guinea pig but growing to 4 ft. in length; aquatic, it has partly webbed feet.

‡Alcide Dessalines d'Orbigny (1802-57), a French naturalist, toured South America 1826-34 and described his experiences in several works.

Tobaco Island, Page shot a monkey and another man got "an enormous stork."[31] Even the waters were abundant with life, and around one hundred fish, averaging two pounds each, jumped into the boat that carried the men back and forth from the *Water Witch* to the shore![32] With such abundance so easily obtained it is not surprising there was little interest in economics. Perhaps this is the reason, too, why the Log records, simply, "McNeil, seaman, deserted."

The steamer got under way again, and as she moved toward Corrientes the river was bounded by rolling banks and great grass plains. Enclosed and cultivated grounds became numerous.[33] Then, near Corrientes, the men saw for the first time the gorgeous *Victoria regia*—the huge water-lily-like plant,* with its immense rose-white flowers and leaves five feet in diameter, "capable of sustaining the weight of a man."[34] They also saw their first Indian encampment, which numbered about 150.[35] At 5:00 p.m. on September 22 the *Water Witch* dropped anchor at Corrientes, the first United States man-of-war to visit the city.[36]

Corrientes† was a large city of 12,000, laid out in the typical Spanish colonial pattern. Although it was already evening, some of the officers went ashore and "walked up and down through unpaved streets and sidewalks well adapted to breaking the necks of people in the dark," commented Page,[37] to whom two bands of Indian musicians "played pretty well."[37] The next day they met the governor of the province, Don Juan Pujol, who was greatly interested in their work and invited them to make expeditions inland whenever they wished.[38] The whole ship was given liberty and both officers and crew saw what they could of Corrientes and its people, including some Indians with "their hair cut similar to the Indians of California."[39] On the twenty-fifth several of the distinguished families of Corrientes visited the ship, the Bedoyas and Billinghursts being most prominent. One young lady "observed archly that we would find '*many animals in Paraguay.*' "[40]

The next day the *Water Witch* once again headed north, now not very far from the confluence of the Paraná and Paraguay rivers and the beginning of Paraguayan territory.‡ The whole upper half of Corrientes had been pampa country of fertile soil, with grass "in luxuriance unequalled," well watered, and having banks studded with timber and wood for fuel.[41] At 4:50 p.m. the ship arrived at Tres Bocas, the first Paraguayan guard station. She sent up the Paraguayan flag and offered a thirteen-gun salute which the Paraguayan flagship answered.§[42] There were four vessels at the station, the largest one of fifty tons and mounting six guns. "The

*This aquatic plant is generally known by its Latin name. Its leaves are 5 to 6 feet in diameter with rims 3 to 5 inches high, green inside, sometimes crimson outside. It is also called "mais del agua" ("water corn"); a bread can be made from its corn-like seeds. It impressed Page considerably:

"What infinite study is found in its leaves—those great pages of Nature's book! I never wearied in examining their mechanism. Here, spreading over the lagoons, they looked as if they would bear the weight of men, and were covered at all times after dawn with myriads of water-fowl, gleaning the 'corn.' . . ." (Page, 265.)

†Corrientes (San Juan de las Siete Corrientes) had been called Taragui by the Guaranís, meaning "lizard," for the countless numbers of them on walls, roofs, and patios. Its Spanish name, "Saint John of the Seven Currents," derived from a point upstream where many eddies seemed to rush in different directions. The Spanish town was founded 1588.

At Corrientes Page forwarded to the Secretary of the Navy a complaint from his engineers, protesting their assignment to improper duty—standing watch as deck officers. (Dobbin took no action.) Page himself was also doubling as the expedition's purser. (Ponko, 121, 246.)

‡The Spaniards' "Paraguay" referred originally to the river; the country's full name is "Republic of the Paraguay." Over 700 miles long, the stream begins in Brazil's Mato Grosso and divides Paraguay in half: the north and west segment is the wild, sparsely inhabited Gran Chaco, and the south and east segment is populated—essentially Asunción and its environs. (Weil et al., 2, 15.)

§When he could, Page refrained from firing salutes, as the shock could damage his delicate chronometers. Despite such care, he soon lost many of his scientific instruments anyway—if not by rank (see n., p. 13), then by wear and tear: the barometer and thermometer were both "accidentally broken"; "our daguerreotype **[sic]** instruments" were also "lost," and "the pocket chronometer had lost its uniform rate. . . ." (Page, 222, 232, 349.)

men were neatly dressed in white trousers and red shirts and were strong, robust looking fellows."[43] At 5:20 the Paraguayan admiral came on board and rode on the ship a short way up the river. When he disembarked the vessel was anchored for the night.

As the ship ascended the Paraguay the following day it seemed to the men as though they had entered an entirely new world, and this world itself was strangely divided. The left bank was a profusion of tropical foliage with broad-leafed bamboos and dark green trees that came down to the water's edge and were twined around with flowering vines which crept everywhere and seemed "to mat the verdure together."[44] The right bank, on the other hand, especially after they had passed the yellowish-red waters of the Bermejo River, seemed sparse in vegetation, often filled with much marshy land from which came swarms of mosquitoes "exceedingly numerous and bloodthirsty."[45] They were to be an unabating annoyance during most of the expedition. This part of Paraguay was the Chaco,* "the home and almost boundless domain of various tribes of inhospitable Indians," Page warned.[46] Many of them, however, proved on several occasions to be quite friendly.

The land was not much inhabited, but every three miles there was a *guardia*† of from six to twelve men who served as lookouts, messengers, and river police, especially against smugglers.[47] This had been part of Dr. Francia's system and López had found it convenient to continue it. At each of these centers were several posts about twelve feet apart, supporting a platform some forty or fifty feet high. This served also as a retreat from the mosquitoes which usually did not fly at that height above ground.[48]

On September 28 the *Water Witch* passed Pilar and anchored that evening off Nembucu, a small town of one-story houses covered with tiles, where, to the mild surprise of those on board, most of the men were dressed in white and the women in red.[49] The next day the ship moved on to Villa Franca, where crowds came to the shore to look at the steamer and to present a bullock to the crew with "all the politeness shown which was possible."[50] By sunset of September 30 the ship had reached Villa Villeta, twenty-five miles from Asunción, where she made her last en route anchorage. In each village there seemed to be something distinctive, and here every house had a small roof-covered porch. As a token of welcome the women lit cigars and offered them to the officers.[51]

October 1 was a memorable day in Asunción.‡ Late in the morning the *Water Witch*, in full dress and with the Paraguayan flag flying from her prow, steamed into the harbor of the city and fired a twenty-one gun salute, which was answered by the shore battery.[52] A large welcoming crowd was on hand, and "an honorary guard of a hundred little boys" followed the officers to the *cabildo*§ for an official reception.[53]

The meeting with President López took place on Monday, October 3. Ammen recorded the scene: "Lopez was a large, heavy man, apparently three-fourths Indian, fleshy, with massive jowls; he sat behind a table arrayed in a gorgeous uniform, a large cocked hat upon his head, with a large ostrich feather upon it. . . ."[54] Page's impression was that López was "highly intelligent, well read, and familiar with the polity of foreign governments." He added, "He talks much, and well." The interview lasted an hour.[55]

The President was extremely pleased with the arrival of the expedition and assured Page and

*The Gran Chaco or Chaco Boreal, some 115,000 square miles of subtropical forests and plains between Paraguay and Bolivia. Long a wilderness for unconverted Indians, it proved to be attractive for cattle ranching (and for potential oil deposits it would be much contested in the twentieth century). "Chaco" is a Quechuan word meaning "abundance of animals." (Weil et al., 19.)

†Guard house; sentinel post.

‡Asunción (Nuestra Señora Santa María de la Asunción, "Our Lady Saint Mary of the Assumption") was founded by the Spaniards in 1537 as the province capital. In 1580 the capital was transferred to Buenos Aires.

§Town hall; municipal council.

his officers that every facility possible would be afforded them in order to insure its successful outcome. Two days later several men from the ship moved into a house on the Calle del Sol, which became both a residence and an office where the work on the various charts could be done.[56]

Asunción was a city of 5000, filled with one-story houses made of large bricks and having from six to ten rooms; many of the roofs were tiled.[57] Among several churches there was one to which no one went:[58]

> The good people rarely allude to this, for a fearful mystery overhangs its sacred precincts; it contained at one time all that was mortal of the Dictator Francia. There he was undoubtedly interred, and a monument erected over his remains. One fair morning the church was opened, as usual, for prayer; this monument was scattered in fragments upon the floor and the bones of the tyrant had disappeared forever—nobody cared how, nobody asked where. It was only whispered that the devil had claimed his own, body and soul.

So fearful was the memory of this man, wrote Page, that "a Paraguayan never mentions the name of the Dictator without looking behind him." Francia's successor had followed in his footsteps, perhaps in a milder way, but his authority was "despotic and unquestioned."[59]

As to the inhabitants, there were "no people more simple in their habits," said Ammen, and "polite beyond measure." "The common character of the Spanish language appears here and everything is *mañana, mañana*, and never *hoy*."*[60]

On October 11 the crew started landing the materials which had been brought for the small steamer and re-supplying the ship, which was being completely overhauled and cleaned.[61] The *Water Witch* was preparing for one of the most intriguing parts of the expedition. The next day the crew landed and set up the howitzer sent by President Pierce; it was fired nine times in honor of President López.†[62]

Among the foreigners in Asunción was an American woman named Mrs. Gillemore. When Page and some officers visited her home she gave him a tame tapir for his zoological collection.‡ The house in which she lived was an old colonial mansion, and Ammen saw "año 1637" carved into one of the window sills.§ Mrs. Gillemore accompanied the officers on a visit to Mrs. López at her villa five miles from the city. Ammen summed up the visit in two sentences: "She has a pleasant manner and is cleverer than her two daughters who excel in being fat. They look harmless and contented."[63]

On October 17, at 6:30 p.m., the Uruguayan boat *Fanny* arrived, bringing Edward Hopkins and his brother, Clement. The officers of the *Water Witch* were on hand to receive him, though the Paraguayan officials were conspicuously absent.[64] This was undoubtedly because Hopkins had already made the first of his many diplomatic blunders when he met Francisco Solano López,‖ who was in Buenos Aires on his way to Europe, and told him the United States had not

*"Tomorrow, tomorrow, and never today."

†Scientists might not have considered the howitzer a suitable expedition gift, but it was another manifestation of modern yankee technology. Page fired it "double-quick time" across the Paraguay River, "keeping two shrapnells [sic] at the same time rolling [skipping] on the water." Page momentarily forgot himself somewhat when he demonstrated the "deadly effect of this modern missile of war," that is, shrapnel (n., p. 60). (Page, 126.)

‡Shy, gentle, nocturnal, the tapir (locally called *aguara*) can grow to the size of a small horse, to which it is distantly related. See also n., p. 30.)

§In the Asunción house where he boarded Hopkins had earlier seen a door sill carved "1632." The sill was hard lapacho wood, still in perfect condition. No doubt Mrs. Gillemore rented rooms. (Hopkins, "Memoir," 29-30; Hutchinson, 250-51.)

‖Solano López was on his way to France to represent his country among the nations of Europe. His father had begun to enjoy the pomp of international attention and quickly discovered the advantages of modern technology. "By the introduction of improvements from industrial nations," said Peterson, "President López had planned a new day for his country. But,

ratified the treaty of the previous March because Paraguay had failed to send a minister to Washington, and that he agreed with the United States in this matter.[65] Correspondence years later proved that López' son had, in fact, relayed the conversation to his father.

Hopkins' commercial endeavors had begun as badly as his diplomatic ones. He had sent a small steamer, *El Paraguay*, from New York on March 21, destined for Asunción with a cargo of hoes, cotton gins, plows, shovels, scythes, and blacksmith's tools—the things he knew from his previous visits that Paraguay needed, but high seas and winds had forced the vessel into Maranhão, where she was pronounced unseaworthy to continue the trip.[66] But Hopkins was determined to continue, and in spite of everything he obtained from López permission to establish a company at San Antonio, eight miles south of Asunción. It was later to be the source of much trouble.*

The October days on the *Water Witch* were filled with little problems and regular routines. Some men broke liberty, only to return after a few days. One of these was Samuel Chaney. William McNeil, who had deserted in Bella Vista, was returned by the *Fanny*. William Baker was disrated for insolence and insubordination. The oil was found unfit for engine use and finally put into the lamps. Small things, these, but they show that the human element and the daily grind were also a part of the expedition.[67]

Much of the time was taken up with the work on the small steamer. By the twentieth the keel had been laid. López had ordered that all the timber necessary should be given, refusing to take any payment for it.[68] He also supplied a number of carpenters who worked from sunrise until 7:00 a.m., from 7:30 to noon, and from 2:00 p.m. until sunset. They were paid from twenty-five to sixty cents per day, depending on their jobs.[69]

The work of the expedition was also proceeding at this time. Daguerrotypes of the Cathedral and the Cabildo were made, two boas nine feet long were captured and added to the collection,† and an interesting encounter—such as an anthropologist would enjoy—took place in the Chaco. This happened at the end of October when one day Ammen, Welsh, Murdaugh, and Stump decided to cross the river and explore the other side. They went three miles up a creek in a small boat and then walked inland for about a mile and a half. Suddenly they met a band of Indians—about fifteen men, their bows and arrows poised, ready for use. They were accompanied by some twenty-five women and children. The explorers waved their handkerchiefs, and the sign was understood.

When the Indians came closer the men from the boat could see that the lobes of the Indians' ears were slit, and a small roll of palm leaf had been inserted to stretch them. Some of the lobes had holes seven inches in circumference. The men were about five feet five inches tall, well developed, "good natured and disposed to laugh and talk."‡ The sailors offered knives and small

undoubtedly, he hoped, too, for personal enrichment. . . ." Like Francia, whom Page called the "principal, or, in fact, the only merchant" in Paraguay, López already possessed over half the land in Paraguay and was enjoying immense profits from yerba maté and timber monopolies. By 1858 he would dominate steam shipping on the country's waterways, with eight vessels in service. Between 1851 and 1860 he trebled Paraguay's exports, and with the profits was financing his national modernization program. (Page, 203, 235; Plá, xvii; Peterson, 257; Brodsky.)

*In addition to the small tools Hopkins had also ordered (sent on another ship) cigar-making equipment, a steam sawmill, sugar mill, and a brick-making machine. At San Antonio, where water power was available, he built the first power sawmill in Paraguay and also started making cigars and bricks. (Peterson, 255-56.)

†Of another boa only seven feet long Page admitted to ". . . some difficulty in putting the huge reptile alive into alcohol without injuring it, as a specimen." (Page, 171.)

‡These may have been Lengua Indians (see n., p. 29), also called "Orejones" ("Ears") because they enlarged their ears with discs. Both "Lengua" and "Orejon" were probably Spanish names for Indians earlier known as "Mascoys." Some reports claimed the Indians never laughed; here Page shows this was not so. "Tigers" (below) and "lions" are commonly the jaguar and puma.

trinkets, and received a few feathers in return. They took the Indians back to the boat and showed them the ship's ensign, indicating it as a flag of peace. As suddenly as they had come, the Indians disappeared into the high grass and trees.

Not long after the Chaco exploration a number of Indians appeared at the river bank and hailed the vessel. Some of the men took beads, oranges, and cigars and went over to meet them. The Indians offered to help the men hunt "tigers," which they said roamed the woods nearby, and they promised to bring all kinds of medicinal plants.[70] Apparently López heard about the meetings, because when Page received permission for the trip up the Paraguay River the men were asked not to have any dealings with the Indians.[71]

On the morning of November 4 the *Water Witch* was rigged in full dress and a salvo of twenty-one guns was fired in honor of the birthday of President López.[72] Lieutenant Page and the officers of the ship were invited to a reception, as were many other guests. The President received one "class" of people at a time, starting with the clergy.[73] When the foreign visitors entered the reception room they were warmly received by López who condescended, for the occasion, to take off his hat. Everyone was afterwards invited to a refreshment room where they found "the most ample spread . . . in very good taste."[74]

Two days later the officers moved the scientific instruments and their personal effects on board the *Water Witch*. Another chapter in the great adventure was about to begin.

Steamer Pilcomayo and Paraguay guardia.

4. THE HEART OF A CONTINENT

If any man ever had a sense of mission, Thomas Jefferson Page certainly did. An extremely purposeful man, he was continually preoccupied with the objectives of the expedition. On the trip to Asunción Page had made it "the established rule of our work to anchor and take observations at all important points."[1] No sooner had he arrived at Asunción than he wrote a letter to the President informing him of the expedition's progress, diplomatically saying, "The intelligence of Your Excellency will readily perceive and as justly appreciate its contributions to Science," asking at the same time for "all the aid and protection it may require."[2]

In Buenos Aires Page had already become aware of the complicated situation in the La Plata region, and he saw that it might make matters difficult for the expedition. The day after his arrival at Asunción he wrote Dobbin:[3]

> The question in dispute between these neighboring Nations, Paraguay, Brazil, the Argentine Confederation and Bolivia relative to boundries [sic] of territory are now in such a State as to induce a most jealous feeling and opposition on the part of Paraguay and Brazil, to the Exploration and Navigation of certain waters, and tributaries of the River Paraguay.

He hoped, however, to persuade the President to let them go ahead, in spite of the fact that the Brazilian Minister—so he thought, at any rate—was trying to prejudice López against the expedition. He wanted, in addition, to explore the Bermejo River, which was closed to all foreign ships because it was a source of dispute between Paraguay and the Argentine Confederation.[4] After being in Paraguay only a short time Page realized that the situation between Paraguay and Brazil was a little more serious than he had at first believed. He told Dobbin, "The present state of affairs, between these two countries, has assumed recently an aspect indicative of war. . . ."[5] He went ahead with his original plans, however, and on October 13 he wrote a letter to López asking for permission "to ascend the river Paraguay, higher up than Assumption, and any of its tributaries which may prove to be navigable," assuring him, lest he hesitate to let a man-of-war steam up the river, that his purpose was "solely on the part of my government to contribute to the promotion of science. . . ."[6]

Some days later Page wrote to the Secretary of State, William L. Marcy, giving him an account of the expedition and the political situation, and telling him that López had somehow found out about Hopkins' statement that the United States would not ratify the March treaty. Page had reassured López that it would be signed, and offered to assist in exchanging the formal ratifications if Marcy wished.*[7]

Perhaps this action helped to appease López. In any event, on November 5 he wrote to Page:[8]

> You are not mistaken in considering me as actuated by the same desire as the President of the United States, to contribute to the advancement of science; and these sentiments, for which I am indebted to you, I think I have sufficiently demonstrated. I renew the assurances that the necessary orders are given in order that every facility and assistance may be extended to you in ascending and descending the river, and wish you the most prosperous journey.

Along with the letter was a passport giving Page permission to ascend to Bahía Negra, where the boundaries of Bolivia and Brazil began, though they were by no means clearly defined at the time. The document stated, "Let him receive all the assistance he may want." Don Feliciano Ramírez was appointed to go along as a guide.†[9]

With all in readiness, Page left Ammen, Lamdin, and Stump to continue the work on the small steamer, and at 5:27 on the morning of November 7, 1853, the *Water Witch* weighed anchor and started upstream. Not far up the river the ship made her first stop, at one of the President's country estates called Lasararaxas—his wife's maiden name. It stretched for six miles along the river.[10] A small port enabled them to anchor and take on wood which was abundant in the area, since both banks had thick jungle almost to the water's edge.[11] Page had already discovered that Paraguayan wood was excellent for fuel, and that "A cord of this wood nearly equals in production of steam to a ton of anthracite coal." In addition, it was cheap, costing only $4.50 a cord, "one third the lowest prices in Montevideo."‡[12] Everywhere they stopped for wood there were plenty of people eager to perform the labor, so much so that Page told Maury, in Washington, "I have instituted one new source of money making to the people—should steamers ever navigate this river—cutting wood."[13]

As had been the case with the lower Paraguay, the upper region also had great variations in vegetation and landscape. The ship passed great groves of palm trees, strange swamps filled with both bushes and trees, and large tracts of marsh grass interspersed with little lagoons.[14] Each of the small rivers emptying into the Paraguay was a temptation to investigation for Page, but he continued on, determining that the Confuso River particularly should be explored some time later on. From time to time they would explore along the banks, and one of their delightful

*López seems to have intimated his knowledge of the treaty's status in a private conversation with Page. Page's assurances seem just as premature as Hopkins' had been, earlier.

†The translation of the letter and passport into English is undoubtedly the work of Edward Bushell, Page's private secretary.

‡Apparently when he returned to Washington Page realized that $4.50 per cord was not cheap at all. Despite the abundance of timber, cutting cord wood was "a new branch of industrial enterprise"—and the natives charged accordingly. (Page, 119.)

discoveries was the delicious fruit of the *guayava blanca*;* wax nodules, deposited on its branches by a species of ant, were also used by the natives to make candles.[15]

A broken crank pin strap caused a day's delay, but on November 12 the *Water Witch* anchored off the town of Concepción,† the largest settlement north of Asunción. The ship's arrival caused tremendous excitement, and the whole next day the vessel was crowded with visitors.[16] That evening a ball was given by the *Comandante*, and even in this relatively remote region "high society" showed up in its best dress. "There is no village or region of the earth so small or remote as not to have its 'upper ten.' " Page reflected.[17] What fascinated the men most was the indefatigable energy with which the townspeople—especially the women—could keep on dancing.‡ Page found himself worn out after a short while and turned his attention to the abundant cakes, red wine, caña,§ and maté.

Among those present at the ball was a Brazilian named Tachiera.‖ Francia had refused to let him leave the country, and he eventually resigned himself and settled here. This old gentleman gave Page a small box filled with the fragrant gum of the *palo santo*, extracted by boiling the wood, and asked him to send it to the president of the United States.[18]

All the settlements the ship had encountered had been on the left bank of the river. The Chaco side seemed completely deserted, almost as if some law had prohibited any habitation there. The first signs of life came when the ship was only thirty miles above Concepción, where a number of Indians of the Lengua¶ tribe were observed. "The men were good-looking, well made, and above ordinary stature, their eyes long, not unlike the Chinese, but larger;" Page noted, "the mouth wide." The women were disfigured by tatooing, which was a preparation for marriage.[19] Later in the afternoon another group of Indians appeared; this time a party from the ship went ashore where they traded cotton twine and wool for tobacco. The *cacique* and another Indian went on board but showed no astonishment at anything. Afterwards the Indians followed the steamer for some distance on horseback, indicating that they wanted more trinkets.[20]

The river was still at least fifteen feet deep, more than ample for the draft of the *Water Witch*. The country around Concepción had been "flat, sandy, and uninteresting," but it now changed to grass plains with palms, then to rolling land forming ten-foot cliffs at the river banks.[21] This was the landscape at Villa Salvador, where the ship arrived on the sixteenth. It was extremely hot—even the water was eighty-six degrees—and no wonder the 1000 people of this village wore very little clothing. Page pithily observed, ". . . many altogether despise this essential of civilization."[22] Scientific curiosity entered again in the form of an interesting species of aloe¶

*The guava; a small tropical tree with very sweet yellow fruit, tasting like strawberry. It is usually found wild; very little is grown commercially. (Page, 133.)

†The Spanish town was founded 1621.

‡Paraguayans were accustomed to dance barefoot; wearing shoes was a formality later introduced by Madame Lynch. Folk dancing and the polka, waltz, and gallop are still popular today, as are indigenous instruments—sugar-cane flutes, gourds, rattles, bells, as well as Spanish guitar and harp. (Weil et al., 94.)

§Caña is rum made from sugar-cane juice; it is the most popular alcoholic beverage in Paraguay. Taken "neat," Page found it "the most potent tipple" he ever tasted. For maté see n., p. 44. (Page, 110.)

‖The word "Tachiera" should probably be "Teixeira," a local name dating to Portuguese explorer Pedro Teixeira, who in 1637 sailed up the Amazon. After a time Tachiera was allowed to leave the country, but he had by then married a native girl—and Paraguayan women were prohibited from leaving the country. The palo santo (below), "holy tree," or guaiacum, guayacan, has a resinous heartwood furnishing an oil used for medicines. (Ponko, 120; Page, 140.)

¶The Lengua ("Tongue") Indians of the Gran Chaco were originally so named because they wore labrets (ornaments inserted through the lower lip). Cacique (below), an Arawak word meaning "lord of vassals," came to be used by the Spaniards for any Indian found in authority; a chief.

¶Aloe describes a liliaceous genus, some species of which produce a strong fiber; it is also known as agave and century plant. Called *caraguatay guazu* by the Guaranís, it was made into rope, fishing line, fabrics, caulking; scraped with a knife the leaves

used locally to make cloth, ropes, and fish lines, so strong was its fiber.[23] This same scientific interest brought their stay to an amusing ending when Page hired a boy to find reptiles:[24]

> I found that I had started a new branch of trade in Salvador, for a goodly portion of the population turned out reptile-hunting; and had I not left the following day, I might unintentionally have worked one of the miracles of Saint Patrick for this neighborhood.

Three days later the *Water Witch* reached the Apá River, an important junction, since from this point onward the left bank of the river interested Brazil, though the territory was undoubtedly Paraguayan at the time.*[25] Page reconnoitered the region for a short while in one of the small boats, and chased a tapir which got away in the thorny thickets. The thermometer now read ninety-six degrees.[26]

The Tropic of Capricorn runs through Paraguay just below Concepción, so the steamer was well into the Torrid Zone by this time. The surroundings had a lush tropical tinge—brilliant flora, birds of gaudy plumage, the strong scent of earth and trees—in a splendid "wild grandeur." Then, to "the admiration of all on board," in back of this wonderful wilderness rose a range of mountains on both sides of the river, as though the ship had suddenly entered a valley.[27] By the next day the jungle had given way to a large plain, broken only by the abrupt rise of a single volcanic peak 1350 feet high. It was called Pan de Azúcar† (the guide probably furnished its name). Page, Powell, Dr. Carter, and the pilot climbed to the top of it to view the surrounding country, and its height was taken and added to the scientific data.[28]

On November 25 the *Water Witch* arrived at Bahía Negra, surrounded by an extensive grass flat—"a sea of grass"—broken by little lagoons and scattered patches of brush and palm trees. The grass was twelve feet high and between a quarter and a half inch thick. In the Bahía Negra River fish were extremely abundant, especially the beautiful *dorado*‡ which was two to three feet long and weighed eighteen to twenty-four pounds.[29] The steamer ascended the Bahía Negra for thirty-six miles, and although the channel was still deep enough, the ship was stopped by an impenetrable growth of grass which seemed to extend all the way to the mountains of Coimbra and Albuquerque, which could be seen forty miles away.[30]

Page was now faced with a delicate problem. He had come as far as his Paraguayan passport permitted, but the Brazilian government would allow him to ascend as far as Albuquerque. In the interests of science and the expedition he decided to continue to Coimbra. He had no idea how López would react to his decision, but it was all too interesting to turn back now. No large vessel, let alone a steamer, had ever come this far. Page wanted the United States to have the honor of being the first nation to navigate to the interior of Brazil. The next day, therefore, the *Water Witch* worked her way back into the Paraguay River and followed it around the bend which takes it northeast into Brazil. At 7:36 on the evening of November 27 the steamer anchored off Fort Coimbra in four and a half fathoms of water.[31]

The fort commander, Antonio Peixoto de Azevedo, welcomed—and was extremely enthusiastic about—the arrival of the ship, calling it "an event worthy of commemoration by the

yielded a sweet liquid; thickened over fire it became a sugar; fermented it made wine, and exposed to the sun, vinegar. The juice also healed wounds. (Page, 145.)

*The left bank of the Paraguay, from the Apá to the Branco rivers, was ceded to Brazil in 1870, after the Paraguayan War; see n., p. 62. (Kolinski.)

†"Sugar loaf"; usually a hemispherical cake of brown sugar produced for household use.

‡Of 238 species of freshwater fish only about 50 are eaten, and these mostly appreciated in urban Asunción. The dorado (a gold-speckled game fish reaching 6 feet in length and weighing 40 pounds) is very popular. Page apparently found a smaller variety. (Wilcocke, 478; Page, 171.)

erection of a lasting monument"—perhaps a marble pillar (fine specimens of marble were added to the collection here)*—and that the *Water Witch* "would live in the memory of the Brazilians."[32] When Page asked him about ascending further, the Commander gave him permission to go to Corumbá, which was slightly farther than the original permission allowed. Page saw the possibility of progressing quite far into Brazil, port by port, and told Maury, "Had the Government of Brazil as much intelligence and liberality as her Comandante at Coimbra, I should find no difficulty."[33] The Commander came on board and rode on the ship for two and a half hours as she steamed toward Albuquerque.[34]

The *Water Witch* anchored at Albuquerque overnight, continuing to Corumbá the next morning. Another pilot who spoke Portuguese had come aboard, bringing the total to three. The steamer almost ran ashore several times while the three men loudly disagreed in Spanish, Portuguese, and Guaraní† as to how she should proceed.[35] On shore there was not a sign of human life, and the scenery was such that Page thought "this mingling of mountain, forest, plain and rock is inexpressibly beautiful."[36] The river was still six hundred yards wide and amply deep. Late in the afternoon the men saw "a remarkable range of mountains" with steep declivities and perpendicular faces. At the highest point, in the center, was a "saddle." They thought it unusual enough to make a Daguerrotype of it. At 5:59 the ship dropped anchor in fiften feet of water at Corumbá.[37]

Though there was little to mark the occasion, it was a historic moment. The *Water Witch* had come 150 miles into Brazil—600 miles farther than any other vessel had ever gone, 700 miles from Asunción and 2000 miles from the ocean. She had traveled to the heart of the interior, to nineteen degrees south latitude and approximately fifty-eight degrees west longitude—almost the very center of the continent.[38]

Corumbá was mainly a collection of thatched huts. In addition to a commander and fifteen soldiers there were about thirty women and children.[39] The place was "simply a military post occupied by Brazil, merely for the purpose of maintaining the right of possession against any claim Bolivia may attempt to assert, to the surrounding territory."[40] During the day it was normally around ninety-six degrees. Mosquitoes were almost intolerable day and night.[41]

Page seriously considered continuing his journey, but his plans were defeated by the very river he was conquering. The stream had begun to fall an inch or two a day,[42] and Page calculated that at its low point it would be about ten feet deep—only a foot more than the *Water Witch*'s draft. To go farther might risk the whole expedition, but Page thought that perhaps the trip might be made later on in the small steamer.

Now that it was certain that the Paraguay was navigable to Corumbá, the effort to reach Cuiabá, the capital of Mato Grosso, was certainly worth while. Page felt a steamer of lesser draft could ascend perhaps another three hundred miles. At present, Cuiabá was sending its minerals and diamonds overland 1200 miles to Rio, a thirty-day trip at best. By water the shipment would take only twenty days, perhaps less. There was plenty of attraction for commerce, because the resources of the area were "as abundant as the bountifulness of Providence had bestowed upon any section of the Globe."[43] To Page this opening up of the interior was the most important and the greatest achievement of the expedition.

On December 3 the *Water Witch* turned her prow south and began the long descent. The first stage was short, for the next day she stopped at Albuquerque where she stayed for five days. The village was situated three miles from the river in a grove of tropical trees and was made up mostly of Indians.[44] It was under the administration of the commander of Coimbra. Page judged the

*Marble and serpentine along the Paraguay River has still not yet been quarried commercially.

†See n., p. 48.

place well run and wrote, "The policy of Brazil towards these Indian tribes is both lenient and humane, and well worthy of imitation of any government. . . ."[45] The real influence in the settlement, however, was the Franciscan missionary stationed there, whose work and efforts brought "respect for the religion and for the order."[46] The padre gave a dinner at his house such as his simple means would allow, and afterwards there was a dance, the favorite way of celebrating any special event. "The western wilds of Brazil never before re-echoed the song and laughter of a merrier party," Page later noted.[47]

Even here in this remote region the *Water Witch* was to have a hand in history again. Four Bolivian officers, refugees from an unsuccessful revolution, had traveled 2200 miles by land and river, getting only as far as Albuquerque. They petitioned Page to take them down the Paraguay as far as possible, even to Asunción, and he readily agreed. These men were General Gregorio Pérez, Colonel Hilarión Ortiz, Lieutenant Colonel Isidoro Reyes, and Dr. Antonio Zaveo.*[48]

One day of the stay at Albuquerque was devoted to a visit of a community of some 5000 Indians. It had been formed within the last ten years by the Franciscans and had all the characteristics of the ancient religious conversions.[49] Then, on December 7, the *Water Witch* got under way once again with several citizens of Albuquerque and the commanders of Corumbá and Coimbra, along with the Bolivian refugees, as guests.[50] In Coimbra the ship's officers, accompanied by several men from the fort and a few others, made a trip to an interesting site called Grotto Inferno. The men descended through a fissure in a rock into a cavern resembling a hugh hall whose roof was supported by twenty- to forty-foot columns. Stalactites which glittered like myriad gems in the torchlight hung from the ceiling. The air was close and about eighty degrees—a factor, perhaps, which gave the place its name.[51]

On the eleventh the steamer sailed from Coimbra, stopping at Salinas, where Page examined some salt flats and a number of Indians visited the vessel, and at Guardia Concluencia, to get permission for the Bolivians to pass down the river. By the fourteenth the steamer was at Villa Salvador where the Bolivians were detained for instructions from Asunción.[52] Page's description of the previous evening could almost be considered symbolic of the expedition he had just about concluded. "Anchored, as the sun was setting in glorious majesty, amid a sea of crimson, gold, blue, rose. How gorgeous are these tropical sunsets, and how solemn, as all nature, with short interval, sinks into shadow of night."[53]

A glorious trip—but there was still the "shadow of night." The first inkling of it came at Concepción, where the Commander was extremely cool in his attitude. When Page tried to arrange a visit to some yerba† fields nearby he was told this was impossible, and a number of dubious reasons were given as to why. Page suspected something was wrong, but could think of no particular reason at that moment.[54] Even the weather was ominous in its indications as lightning and rain accompanied the loading of wood onto the vessel for the next two days. With the current in her favor the *Water Witch* made the trip from Concepción in three days, arriving at Asunción on December 20.

Page was scarcely back in the city before he received a note from Benito Varela, Minister of Foreign Affairs, asking him how far he had gone on his river penetration. Page recalled the attitude of the commander at Concepción, and sized up the situation at once. He determined, however, to answer the minister frankly, and on the twenty-first sent him a note saying he had gone as far as Corumbá, adding, perhaps in the hopes of getting a more sympathetic hearing, that he had been treated with kindness everywhere.[55] Three days later the official newspaper,

*The political candidate of these officers actually came into power in Bolivia in 1857; see n., p. 74.

†On yerba maté see n., p. 44.

Seminario, carried an article about Page's trip, stressing the fact that he had violated his passport.[56]

Surprisingly, it was Hopkins who came to Page's defense. In a report to Secretary of State Marcy he said that Page had misunderstood the permission given him by President López "as to the terminus of his late trip to the northern frontier," and that he himself had assured the President that the expedition would have the most brilliant result. He was certain the coolness on the government's part would change, adding that the President had shown "every favor" to the United States and Paraguay Navigation Company.[57]

Page considered the whole thing "child's play and absurd,"* and argued that López had no authority to prohibit him from going into Brazil if he wished to do so.[58] He even argued that López had said nothing about Bahía Negra being the limit, officially, and that the whole problem was merely the President's "suspicious and irritable temper."[59] Had Page put himself in Paraguay's position he might have better appreciated her fears about opening the interior. His own determination, which was close to stubborn insistence, to go ahead irrespective of any other considerations was really the cause of the present hard feelings. Nevertheless everything was resolved by Page's giving López a sketch of the upper river to Corumbá, thereby adding to his knowledge of his own country and allaying his suspicions that the ship was working for interests possibly detrimental to those of Paraguay.[60]

During December the final work was being done on the small steamer, which Page decided to name *Pilcomayo*,† after the river which empties into the Paraguay just opposite Asunción. The Christmas season proved a troublesome time; many of the men broke liberty, and two deserted. One of them was Samuel Chaney, who returned after nineteen days.[61] Meanwhile, immediate repairs were being made on the *Water Witch*. Page decided to kill two birds with one stone by sending the first shipment of specimens to Montevideo, where the *Water Witch* could undergo major repairs and take on new supplies. Lieutenant Donaldson was detached from the steamer in order to accompany the specimens to the United States where, according to a previous arrangement with former Secretary Kennedy, they were to be sent to the Smithsonian Institution and labeled as the work of the *Water Witch*.

Page suggested the live animals be put in the Washington zoo. They were all from Paraguay and included a capybara, a *cuati*‡ (coatimundi), a *micuré* (kinkajou), a tapir, and a young deer. There were also five birds: four *motus* and one *iaha*. The fish and reptiles, preserved in alcohol, included crabs, eels, turtles, toads, iguanas, alligators, lizards, two large worms, and a frog with

*Page did not notice that their roles were comically reversed since the flag-flying episode (n., p. 17): Page was now in difficulty and Hopkins was calmly trying to set things right. Hopkins had first praised López as "a man possessing a high degree of talent, . . . character, . . . tenacity . . ." and then damned him as a raging tyrant. Page had found López intelligent, knowledgeable, but now (a few lines below) saw him—guardedly—as "suspicious," "irritable." Back in Washington Page called him "an accomplished, but . . . unscrupulous diplomatist." Each in turn thought the protocol "absurd." The interests and personalities of these men were strikingly similar; see also n., p. 14. (Peterson, 256; Hopkins, "Memorial," 31; Page, 117.)

†Named after the Pilcomayo ("River of Sparrows"), the border between Paraguay and Argentina. Ironically this stream was not explored: it is sluggish, winding, shallow—unnavigable. It was nevertheless on Page's agenda for the second expedition (see n., p. 86). (Ponko, 128; Wilcocke, 69; Weil et al., 16; Kolinski.)

‡The coatimundi (cuati, quaty) is related to the racoon. The micuré or micuna is the kinkajou, a quadruped with prehensile tail, also in the racoon family. Although the crew tried, it never caught a tapir (n., p. 24) in the wild. Among three kinds of South American deer the rare pampas deer, reddish-brown with small antlers, has musk glands whose odor can be detected a mile away; the crew caught one (and failing to remove the gland immediately) found its meat reeking of musk. Page put the motu or mitú in the partridge family but it is a genus of the curassow; the iaha or jaja, taha, is the South American weaverbird. For capybara see n., p. 21.

white eyes. Many of these creatures had never before been seen outside South America.* The collection was completed by several botanical specimens.[62]

During Page's absence the work on the *Pilcomayo* had its ups and downs continually, including a serious quarrel between Ammen and Lamdin over an engineering problem which led them to write long letters to Page to explain the matter, and to vice-consul Clement Hopkins to settle it. Even though Stump testified that "Lieut. Ammen did contradict Mr. Lamdin several times, in the most positive and emphatic manner," Ammen, as an officer, eventually had his way.[63] The quarrel, however, undoubtedly contributed to Ammen's withdrawal from the expedition a few months later.

In addition to this problem the boat was "much strained by a miserable attempt on the part of the carpenter to launch her," and that, too soon. The workmen were slow: "three could do what eighteen are doing" in the same time, Ammen wrote.[64] When it was finally finished the vessel was sixty-five feet long, flat-bottomed, with a three-foot hold and a twenty-three inch draft. A deck house five feet high ran most of the length of the ship. It was made of cedar and covered with canvas. The dining table served as a drawing board and the seats were lockers.[65] On January 7 she had her first test run, and after a few adjustments and trials she was ready.† Ammunition and supplies were transferred to her hold, and a crew of about a dozen men assigned to her. These and most of the officers then watched the *Water Witch* leave for La Plata on January 30.[66]

*Page hoped his collection of live animals would be the nucleus of a national zoo. Congress, however, somewhat tired of science (see n., p. 79), appropriated no money for such a facility. The donated tapir died enroute, but a "tiger" which reached Washington ended up at the Insane Asylum farm.

(Page, 95, 160-61, 195-96; Harrison, 198; Ponko, 117, 132.)

†The *Pilcomayo* had two steam engines and 12-foot paddle wheels. Ammen said foaming of the boiler was what gave it constant trouble. (Ponko, 272; Page, 246; Ammen, 260.)

Port Coimbra.

5. INLAND AND THREE RIVERS

Except for the seven men who had been left in La Paz and the few who were working on the *Pilcomayo*, the expedition had thus far operated as a single unit. With the departure of the *Water Witch*, the work of the expedition fell to those left behind, and Page decided to divide the men into two groups: one which would survey small rivers and streams in the steamer, and the other which would explore the interior of the country. Page decided to head the inland expedition himself, and, along with his son John, he chose Ammen, Powell, Henry, Welsh, and his secretary to accompany him. A Mr. Faccioli, "a young Italian wishing to see the country," also went along. The party was completed by a guide named Juan Talavera and two servants.[1] Page assigned the rest of the men to the *Pilcomayo*, which he put under Murdaugh's command, with orders to explore the Confuso and any adjoining rivers.

When Page applied for the President's permission to make a trip through the country (and he couldn't have done so without it), López gave him a special passport, saying that while it was probably not necessary, any official "would understand nothing you could say so well as a passport."[2] At 3:00 p.m. on February 6, 1854, with their belongings and equipment on three pack horses, the party left Asunción. They traveled six leagues* that day, and spent the night at the house of Don Jaime Corvalón.[3] The next day they rode through open plains and beautiful woods, arriving at Paraguarí around one in the afternoon.[4]

*The old Spanish league varied between 2½ and 3½ miles.

> We dined here and after the usual manner in Paraguay; that is some large earthen dishes in the centre of the table, a horn spoon for each person and sometimes a knife, but this the traveller is supposed to be able to furnish himself. Usually roast ribs of beef come first, then a stew of beef or chicken accompanied perhaps with mandioco,* and never bread, but sometimes by parched corn; then follows a dessert of cheese and molasses all eaten after the manner of the patriarchs out of the central dishes.

That night they slept "gaucho† fashion" with the saddle for a pillow and the horse blankets for bedding.

Two more days of riding brought the travelers to Villarrica.‡ They had crossed several streams of fairly strong current and passed through tall-treed forests spaced as though they were purposely planned to relieve the traveler from the heat of the day. The sun was just setting on the broad plain when the riders entered the town. A man named Don Luis Homen found them a house at which some of the party stayed for several days.[5]

Three separate incidents during the stay at Villarrica gave an excellent idea of different phases of contemporaneous social life. One evening the men were at the home of an Italian who had been living there nine years. The conversation was about dyes and gums—the natural products of a tropical forest, which eventually led to the question of wealth. The Italian remarked that the easiest thing to get for money now were women, and added ironically, "It's the effect of civilization."

The second impression was brought about by the attendance of Ammen and Page's secretary at Catholic Mass. "The service here is like in all countries§ where one religion obtains, a mere matter of form; and how different from a religion of sentiment or of real practical belief." That same day, when he was looking for assistance, Ammen discovered it was the women who were willing to work, and that the men "regard the title of 'Don' as incompatible with any labor whatever and the only exertion compatible with their exalted worth is standing behind a counter and vending cotton handkerchiefs, iron pots, crockery, pins, large iron spurs and other simple wants of this people."[6]

As always, the scientific purpose of the expedition was uppermost in the minds of the travelers. Lieutenants Powell and Henry, with sextant, pocket chronometer, and artificial horizon found the latitude and longitude of all important locations.[7] Ammen was interested mainly in hunting for animal specimens. He spent four days in the wilderness with an old tiger hunter named Miguel Coriaga, but they found nothing. He finally bought two armadillos, which he skinned.[8] Ammen had also taken along numerous items for Daguerrotyping, and made several exposures.

On February 19 Ammen left Villarrica and started back to Asunción. The return route was different from the one by which he had come, and he was much impressed with the scenery. "The appearance of the country is very beautiful; mountains of some 1500 or 2000 feet rising

*Sic—mandioca (the local spelling) or manioc is also called cassava, yuca, and tapioca; it includes several plants with starchy rootstocks from which a flour can be made for bread or which can be eaten cooked. It is Paraguay's largest carbohydrate crop.

†Gaucho (Quechua, orphan, "poor" person), a cowboy or herdsman on the pampas, usually of mixed Spanish and Indian blood. Superb horsemen, gauchos used the lasso (a pleated hide rope 60 feet long) and bolas (three wood or stone balls 3 inches in diameter, each on 3-foot thongs, connected) to capture cattle and other animals. (Page, 52n.)

‡Ammen uses "Villa Rica" in his notes; today it is Villarrica, "Rich Town." It had been founded 1570. The spelling "Homen" (below) is questionable. Page has "Homan" and Powell "Homem" in their notes; since I am following Ammen I give his spelling.

From the three sources mentioned it is evident the party split up here, Page continuing south, and the others returning to Asunción. It would seem that Ammen, Powell, and the secretary returned. Since Page certainly did not travel on alone, it can be presumed Henry and/or Welsh went with him. I have combined Page's brief notes with those of Powell in the next chapter.

§Ammen, probably inadvertently, wrote "in all country" in his notebook. "Don" (below) is a Spanish term of respect, like the English "Mr."

abruptly, covered with large forest trees, narrowing sometimes to a valley of 5 or 6 miles then expanding to 15 or 20."[9] He stopped at several estancias, on his way and found, as Page put it later, there is "no trait of character more characteristic of the people of Paraguay than their hospitality."[10] He arrived in Asunción at 10:00 p.m. on February 23. After another day the *Pilcomayo* returned from her expedition.[11]

At the same time López gave Page permission to go inland he also allowed Murdaugh to examine any of the rivers and rivulets within twenty leagues of Asunción. With this purpose in mind Murdaugh got under way at noon on February 10 with sixteen men aboard the *Pilcomayo.*[12] They passed a few rivulets which, although the current was strong and there were many rocks, had "plenty of water for ordinary vessels." What Murdaugh most wanted to find was a navigable waterway into the Chaco country. He asked various natives he met if any such thing existed, and recorded that "the people on shore here seem to know nothing of any river running into the Chaco. Some of them speak of a small stream coming into the Laguna Nañanao which has a little current, but the accounts are so vague that I place no sort of confidence in them."[13] Finally on February 14 he found the mouth of the Confuso and started up the river.

The steamer was working "remarkably well," but the river presented many problems. Only twenty yards wide and extremely tortuous, it was filled with short turns and obstructions. The banks were so high the men aboard the steamer could not see over them. When they went ashore to investigate, they found a palm-filled grassland and saw the remains of houses in an uninhabited Indian village.[14] It was not for lack of food that it had been deserted, for there were "many seals about," and even some alligators, "the tails of which some of us thought very good eating." The water, however, had "a sweet and slightly nauseous taste." The main problem was "thousands of mosquitoes during the night."[15]

The next day's journey proved the river to be "nothing but turns," and the overhanging trees had become a menace to the superstructure of the ship. Murdaugh decided to turn back downstream. He had found the river to be generally nine feet deep and with a current in some places of two knots.* He concluded, "A river with such a current and so many bends must come from a long distance." Later exploration proved he was right.[16]

A day was spent cleaning the ship, and then the little steamer entered a large lagoon into which the Salado River empties. Although the river was twelve yards wide and twelve feet deep, the *Pilcomayo* could not go far because of overhanging trees and abundant *camalotes*,† a type of water lily with lilac-colored flowers whose twenty-foot stems intertwined in the water, making an almost impassable mesh.[17] Murdaugh decided to explore farther upstream in the small boat, and set out with the doctor and four men to do so. Because they were on the left side of the Paraguay River there was little danger from Indians. They found a little village called Salinas de Salado, where they got some specimens of salt. Here they heard of a very large lake some distance away and determined to investigate. After a night in the open with "mosquitoes, a heavy dew and cool south winds," the men walked nine miles to the lake, which they found to be very shallow but "a very pretty sheet of water." They stayed that night at a nearby estancia where, they learned, a "lion" had killed twenty-two sheep a few nights before. The exploration ended with a long hike and another cleaning of the steamer, which headed back to Asunción on February 24, arriving the next day.[18]

*The nautical "knot" is understood to mean "miles per hour."

†This hyacinth-like plant can sometimes form floating islands. For "lion" (below), see n., p. 25. (Kolinski.)

As February drew to a close the Carnival days came. Whatever the men might have expected, they found very little excitement. Ammen wrote that "the only fun appears to be to dash a cup of water upon the passer by."[19] The spiritual significance of the season seemed to mean more to these simple people. Their attitude was reflected in a custom the men witnessed every day. "At sunset all the church bells rang out with the Angelus; every human being stopped, bowed his head, and said his prayers, whether in mid-street or in his house."[20]

While the men awaited Page's return they were occupied "in arranging the thousand small things that have to be done yet appear to be nothing," said Ammen. He had a list of twenty-one chores, including such items as making wedges, fixing a gun strap, painting boats, shelves for books. One man was making a pattern for a steam drum, and another was building a forge.[21]

In the midst of this activity Page returned from his trip. He had gone all the way to Encarnación (which he mistakenly thought was the boundary between Paraguay, Argentina, and Brazil) and had traveled approximately 600 miles.*[22] In a long letter to Dobbin he gave his impressions of his journey.

The country had "scenery in most of its parts, exceedingly beautiful; and soil fertile as nature could have made it, and of varied quality." Cultivated land was devoted mainly to corn, tobacco,† and manioc. Timber and yerba maté were the most valuable products, but these were monopolies of the government. The government, in fact, owned most of the land, though some private sales had been made, the buyer paying a small amount annually. Most of the time the landowner raised cattle and sheep, as it was "not to his interest to develop the resources of his land, or add to its value by the introduction of machinery; because it may be taken from him at any time."

In addition to this reasoning, the people were not much given to hard work. Page added that "nothing will so fully illustrate the indolence of the people and their aversion to labor" than the fact that they did not even try to raise cotton, "in spite of the fact that cotton in the seed commands 12½ cents a pound."[23] Page had previously written to Dobbin that "There is a fine field for the exercise of American skill and enterprise in this country,"[24] and after his trip he was more convinced than ever. "The country of Paraguay presents a field for enterprise of which other portions of the world are profoundly ignorant."[25]

A few days after Page's return an important event took place. Page informed Dobbin, "The Congress of this government will assemble tomorrow. The most important business to be transacted will be the Election of a President. There is no doubt of the reelection of the Present incumbent, inasmuch as no one would presume to oppose his Election."[26] On March 14 López was elected to another ten year term. The President gave a lengthy message to the Congress and then adjoined it *sine die.*[27] The following week Page and his officers called upon the President to congratulate him on his reelection. They were accompanied by Hopkins, who afterward invited the men to his house for a glass of wine. It was the first such gesture on his part.[28]

At about the same time Ammen coincidentally came across shades of the earlier dictator. He had been invited to take some pictures of the Machaíns, a prominent family in Asunción. He was surprised to learn that "among them [was] an old woman who had been the '*flame*' of Francia.

*Page was mistaken about the boundary, because at this time the Misiones territory belonged to Paraguay. The actual juncture of the three countries was sixty or seventy miles farther south. See also n., p. 46.

†Tobacco was long known to Paraguayan Indians and smoked by women and children in some places, as well as men. Until Hopkins' "American Company" there was no designated district—even in Asunción—for making and selling cigars, although it was an enormous cottage industry. Because of its fine quality Paraguayan tobacco is the country's principal export. Both coffee and cotton (below), also important crops, were introduced by the Jesuits. (Page, 218.)

Her marrying a Machaín brought fines, imprisonments and finally the death of her husband; he was shot after a long imprisonment."*[29]

While Ammen was working with his Daguerrotypes, other members of the expedition were journeying into the country for more specimens, and Powell was measuring the volume of water passing by Asunción.[30] On April 7 Page sent Ammen an order to outfit the *Pilcomayo* and explore the Jejui River.

On April 11 Ammen and his crew, with Mr. Faccioli, whom Page had given permission to go along, set out for the Jejui River, which empties into the Paraguay about a hundred miles north of Asunción. Progress was slow, because the wheels were making only twenty revolutions per minute. By evening they had gone only as far as the Salado.[31] The mouth of the Paraguarí was reached the following day, and the steamer anchored off the estancia of the President's son. A little above Rosario† Ammen intended to investigate a rivulet called Yetete, but was told by a canoeist that it was blocked up, so he went on to the Jejui, entering it on the fifteenth.[32] Although it had nine feet of water, the current was strong and there were many sand bars and short bends. Ammen decided that the river was not fit for vessel navigation, but went as far as the anchorage of San Pedro,[33] about fifteen miles upstream. It proved an interesting experience.

The appearance of a fire-breathing dragon could have caused no more excitement than the coming of the steamer. Dozens of natives, robust people "with no clothing but a hat,"‡ gathered on the bank and stared at the ship and even more at the men, muttering "Que gente!" The commander of the "port" gave the men horses to ride the two and a half miles to the village of San Pedro, where they met the local magistrate, the priest, and ten or twelve others who all returned with the men to visit the vessel.[34] The next day the steamer was so overrun with visitors the men couldn't even eat their dinner. They made up for it that evening, however, when the commander gave a grand party at which "all the authorities and grandees were overwhelming in their politeness," and "the ladies all wore shoes."[35] The celebration lasted all night, with English ale being the main attraction,§ and though it was plentiful Ammen noted that "drunkenness is unknown in Paraguay axcept among foreigners who go there."[36]

At 9:30 the next morning a group of officials and their wives went on board the vessel and had a little excursion of about a mile and a half. Later someone offered the men some milk (asking two *reales*‖ for it), but nobody drank any—it had been brought in a chamber pot![37]

Ammen attempted to ascend the river in a small boat to see if it might be navigable farther on, but he met a continuous series of sand bars. On the twenty-second the *Pilcomayo* left San Pedro, and the following day at 3:35 p.m. she was back in Asunción.[38]

On April 25 the *Water Witch* returned to Asunción after an absence of almost three months. There had been much trouble with the paddle wheels, and great difficulty in adjusting and

*In modern Paraguayan accounts Estanislao Machaín had been guilty in an 1820 conspiracy against Francia. (Kolinski.)

†Rosario should not be confused with the large town of the same name in Argentina.

‡Climate and native indifference to Spanish styles resulted in some Paraguayans wearing almost nothing. But Francia had required all adults in Asunción to wear hats—if only the brim—so as to doff them when officials, even soldiers, passed by. Soon in all major cities it was common, said Brodsky, "to behold a Guaraní walking the streets dressed only in hatbrim and ankle spurs." Their exclamation (below) was "What [incredible] people!" (Brodsky, 42.)

§At this party *tiste* was also served—equal parts of parched corn and sugar dissolved in cold water. (Ammen, 261.)

‖The real in colonial times traditionally was one-eighth of a peso, 1 ounce of silver. Page found a real to be worth 13 cents, U.S. In 1943 Paraguay replaced the peso with the "guaraní," approximately equal to the U.S. dollar. (Page, 360; Kolinski.)

repairing them. Page's original enthusiasm cooled considerably. The wheel was, after all, an experiment, and a lengthy statement sent to Dobbin summarizes Page's judgment of it:[39]

> Often have I wished that the long tried "radial wheel" had been applied to her, instead of the "Morgan Excentric." The constant liability to derangement, incident to the latter; its ponderous weight; its complicated structure, requiring the greatest degree of accuracy in the fitting of all its parts, the friction and the consequent abrasion of the bushings, produced by the slightest derangement; finally, the great expense in the repairs, have all had the effect to produce abandonment of this wheel by both the English and the French.

Lieutenant Moore, who had been in charge of the vessel the preceding months, asked to be replaced on board because he was not well. Page consented and told Ammen to take his place. Whether this move influenced his decision is hard to say, but two days later Ammen asked to be transferred to the Brazil squadron and on May 14 was "relieved from duty on the Expedition."[40]

Inasmuch as Ammen was a keen observer, familiar with Daguerrotyping, and apparently a fairly good scientist, his departure from the expedition was definitely a loss. It is hard to say what his real motives were for leaving. The quarrel with Lamdin has already been mentioned. Perhaps, too, he may have felt personally the fact that the *Pilcomayo*, which was mainly his work, was not functioning very efficiently. Or again, it may simply have been the human element, for although in his letter of request for transfer he stated that he had nothing against the captain of the *Water Witch*, he later wrote, "The captain of the *Water Witch* was entirely a gentleman, but, so far as my opinion is of value, was not well fitted to command such an expedition."[41] Whatever his reasons for withdrawal may have been, it must be said that while he was with the expedition his contribution was considerable.*

Page, meanwhile, felt there were still some explorations to be made in order to render the expedition as comprehensive as possible. He was especially interested in surveying the Bermejo River, the disputed boundary between Paraguay and Argentina, and he wanted a scientific survey made of the territory he himself had traveled. Argentina attracted his attention too, but first he wanted to be able to tell the world all about this land of "sunlit plains, perennial streams," and "budding resources."[42]

*Ammen was assigned to the U.S. brig *Bainbridge*, whose eccentric captain (whom he gallantly does not identify—but see n., p. 65) became something of an ordeal. Ammen found this the worst duty he ever had; when they returned to the United States, he said, "it was my intention to resign . . . rather than be subjected again to so horrible a life." Providently, he was quickly transferred to—of all places—the Washington Naval Observatory, under Maury. (Ammen, 289.)

Curing yerba.

6. SUNLIT PLAINS AND PERENNIAL STREAMS

From the time she left Asunción until her return at the beginning of May, the *Water Witch* was not immediately involved in the work of the expedition. Her role, nevertheless, was important insofar as she had gone to gather necessary supplies and be put in condition for further activity on the Paraná. Even while she was in La Plata waters she had her own noteworthy incidents, and was not without her particular problems. The very day she arrived in Buenos Aires, February 9, four men stole a boat from her decks and deserted. One of them was McNeil, who had previously jumped ship in Bella Vista. The boat was brought back the next day, but the men were seen no more.[1]

On February 12 Lieutenant Donaldson boarded the American ship *Roman* to take the first shipment of specimens back to the United States. At the same time Pendleton came aboard the *Water Witch* as a passenger to Montevideo.[2] The steamer stayed in this port from February 13 to April 14. During these two months the ship was painted, caulked, and re-rigged, and the wheels repaired with special equipment brought from New York for the purpose.[3] Twelve new men were taken on board, including Charles Turner and Charles Johnson, two of the three men who were to die on the expedition, and John Bowditch, who was soon to be one of four on an interesting exploration. Three men were discharged, one as a prisoner to the flagship *Savannah* for having struck an officer.[4]

Shortly after the arrival in Montevideo Commodore Salter,* the squadron commander for the area, came aboard the *Water Witch* for an official visit. His interest in the expedition was genuine, and he wished to make sure that the steamer was getting all that she needed. Through his

*See n., p. 63.

kindness the ship was able to load a great quantity of goods, and a survey of them makes an intriguing inventory. Some of the items were pots, pans, putty, glass, chalk, potash, rope, padlocks, buckets and brooms, and a food supply including forty-one pounds of pickles, vinegar, sugar and beans, and 6106 pounds of bread![5]

On three occasions during her stay the *Water Witch* put on formal dress. The first was February 22, George Washington's birthday. Moore recorded, "All the foreign men of war in port (*except the French*)* and the Fort on shore hailed the American flag."[6] The two other occasions were in honor of Brazil: on March 14, for the birthday of the Empress,† and on March 25, for the anniversary of the ratification of the Constitution.[7]

The American consul, Mr. Hamilton, and his family came on board April 13, and the *Water Witch* sailed for Buenos Aires. On the sixteenth she started for the Paraná River and stopped briefly at all the major points previously visited. She met the Paraguayan ship *Tebicuarí* on April 29 and asked for information as to the whereabouts of the *Pilcomayo*, but the Paraguayans knew nothing. The inquiry was answered two days later when she entered Asunción with a formal salute and saw the little steamer anchored there.[8]

The first two weeks of May were spent in routine duties and in preparing the *Pilcomayo* for another trip. It was during this time that the first tragedy of the expedition occurred. At 4:30 on the afternoon of May 11, Charles Johnson, trying to get from the cutter to the ship, fell overboard and drowned.[9] This sad event was shortly followed by the departure of Ammen. Both events were soon forgotten in the hustle of last-minute arrangements on the small steamer. Page had decided to take most of the officers and the necessary crew to explore the Bermejo River. Powell was left behind with instructions to retrace the trip Page had previously made, with a view to making scientific observations and collecting whatever useful specimens he could. On the morning of May 18 the men left on the *Water Witch* gave a seven-gun salute and three cheers to the *Pilcomayo* as she steamed out into the Paraguay River.[10]

Four days after leaving Asunción the *Pilcomayo* entered the mouth of the Bermejo.‡ There were two accounts of earlier explorations on this river, one in 1790 and the other in 1826, but both had been descents, and in small boats. The trip of the *Pilcomayo* was the first known attempt at ascent.[11] The steamer was not functioning too well, and a three-mile-an-hour current made the going slow. The river varied from one hundred to three hundred yards in width, and was three to ten feet deep.[12] The banks during the first days were covered with grass and palm trees "extraordinary in size and beauty."[13] There were, however, few birds, and very little other animal life. The Indians of the area were constantly killing them off. Ants were abundant, though, some of the ant hills three to four feet high.[14]

Not until they were two weeks up the river did the men on the steamer see any Indians. Those they met were of the Toba§ tribe that frequently communicated with Corrientes. This was

*As there were no international questions or disputes between France and the United States at this time, the refusal of a salutation must have been an arbitrary decision of the French captain(s). The earlier French and British cooperation, however, had also come to an end.

†Brazil's emperors were a striking contrast to the petty dictators of its smaller neighbors. Pedro I (1798-1834), son of the king of Portugal, fled to Brazil in 1807 to escape the Napoleonic Wars and, declaring Brazil's independence from Portugal in 1822, became its first emperor. His son Pedro II (1825-91) aided Urquiza against Rosas and joined Argentina and Uruguay against Paraguay. A man of wide culture, he abolished slavery and encouraged great social progress in Brazil. His empress was Thereza Christina, a Sicilian princess but not a beautiful woman; she was warm and loyal, nevertheless, and Pedro and Brazil both loved her.

‡Bermejo (or Vermejo) means "Red River," so colored by its soil content.

§Members of the Guairucu people of Argentina and the Gran Chaco. The Tobas were the most numerous of the Chaco Indians. "Unitarians" (below) were an anti-Rosas faction which he effectively stopped. The fur nutria comes from the coypu, a large South American aquatic rodent. (Page, 253.)

comically obvious from the clothes they wore, which were those of the Rosas period, including the conical red cloth cap, eighteen inches high, with a brass plate reading (in Spanish), "Death to the savage Unitarians!" The explorers visited a Toba encampment some days later. It contained only twenty people living in huts of grass and hide shreads. Some of the Indians traded nutria skins for cigars.[15] Page later wrote to Dobbin,[16]

> They possess nothing of that interesting character belonging to the Indians of North America: their condition seems to be one of want; their huts, if such they may be called, of grass and matting, afford but little protection. I frequently held a talk with them, making them presents of tobacco, trinkets, etc. They live by the chase, fishing, on roots and such wild fruits as the forest produce. [sic]

When the *Pilcomayo* had ascended the river some fifty or sixty miles the banks began to rise considerably in height, varying from fifteen to forty feet. A small tributary attracted Page's attention, and he, the doctor, and two other men set out to follow it for some distance. As they walked on, the landscape became "a scene of rich beauty." The little rivulet was full of rippling cascades, and the air was heavy with the heady perfume of acacias in full bloom.[17] The four men wandered farther inland and after a while realized they were lost. Night came and they were still trying to find their way back to the boat. The thought of Indians and jaguars pushed their feet faster. Several times they fired shots but received no response. When they finally found the river, the *Pilcomayo* was gone. More shots. A faint sound. An echo? A response? the sounds grew gradually louder as the hikers moved south. and then the men saw the welcome blinking of lantern lights in the darkness. The boat had moved downstream to better anchorage in case a party would have to search for the wanderers. Page later said his feeling of unpreparedness was the worst part of the experience.[18]

The boat continued on the Bermejo until it had gone about 125 miles,* a distance, Page felt, sufficiently far to prove that the river was navigable in a vessel of shallow draft.[19] On July 5 the *Pilcomayo* arrived at Asunción.[20] Another phase of the expedition had been successfully completed.

By the time the Bermejo expedition returned Lieutenant Powell and his companions were far into the interior of Paraguay. Powell, John Bowditch, Henry Kleiz, and a guide and Guaraní interpreter named Roque Ferrida left Asunción at 10:00 a.m. on May 31. They took two tents and several scientific instruments and equipment for preparing specimens. They went through Luque to Itagua where, two days later, Powell prepared a box of fourteen birds and sent it back to Asunción with a man going that way.[21] Across sunlit plains with frequent fields of tobacco growing in the rich, dark loam, the party rode on, reaching Atira, "a collection of from 100 to 150 badly constructed huts with an occasional house supporting a tile roof," on June 8. It was typical of many they were to see in the weeks to come. On the way to Caraguatay Powell saw several unusual insects and added them to his collection. Caraguatay had 120 houses with "the same number of huts used as kitchens." Only the church had a tile roof. On Sunday morning everyone disappeared; Powell discovered later they were all at the local race track!†[22]

The journey turned northward and the riders passed over plains cut up by lagoons and filled with wild orange trees‡ loaded with fruit. Along their way they stopped at the estancia Ygazaum

*The distance here is a compromise between two figures: Page says 122 miles (Page, 248), and his July 12 letter says 130.

†Horses and horsemanship have always been much admired in Paraguay, and fiestas invariably include races. Local raceways are usually two straight, parallel tracks, 400 to 500 yards long.

‡Oranges were probably brought to Paraguay by missionaries, and many trees have reproduced in the wild for centuries. "Bitter" oranges now both grow wild and in groves; the fruit, leaves, and twigs yield petitgrain oil, used for perfumes and

and learned that it was twenty miles square and had 80,000 head of cattle and horses. The next day when the travelers stopped near Arroyo Hondo the magistrate sent word to Powell that he would like to see him. Powell, to the amazement of the natives, sent word back that he couldn't come, but that he would be happy to receive the magistrate at his camp. To the even greater surprise of the people he came, bringing with him a special guide to help them through the marsh country over which they would have to pass shortly. The guide proved helpful the next day getting across country ribboned by dozens of swift little streams. This land changed to rolling country scattered with palms. On June 17 the riders reached San Estanislao and were comfortably housed by the head of the village, José Ignacio Jesús Ricaldi.[23]

Powell and his companions spent several days in San Estanislao. They had arrived just in time to perform a very useful service for the town officials. The President had ordered that the location and an exact plan of the town be forwarded to Asunción, and they were wondering how to obtain the information. Powell gave them a copy of his own observations, which they happily sent off to the capital.[24] The scientific data so gathered also extended to other fields. A mountain turkey and two large red cockatoos were captured. Samples of medicinal plants "such as appeared to be considered of most value by the *Curandero** or Medicine man of the Village," were gathered.[25] Most important of all, Powell had a first-hand look at the mainspring of Paraguay's economic system—a yerba plantation.

The *yerbale*† "Santa Rosa," under the supervision of Señor Fulgencio Parieda, was about twenty-five miles from the village. Powell learned that the collectors and curers of the plant moved from plantation to plantation accordingly as the trees were ready for "harvesting." Preferred trees were from three to four years old, and eight to ten feet high. A supervisor and field workers would build a few houses and a large shed fifty to sixty feet long. The workers gathered leaves brought them to the shed, carrying at times as much as fourteen *arrobas* (350 pounds), for which they received twenty-five cents per load. The leaves were then toasted from thirty-six to forty-eight hours after they had been spread out to dry for some time. The drying leaves were constantly watched so that none of them would be taken. The whole process was under the complete control of the government.[26]

Life in San Estanislao also had its amusements and festive occasions. On June 24 Powell recorded:[27]

> Being the feast of St. John the country people are in [town] and at night we had a ball. I was the *Beau* of that ball "tho' I say it, who ought not to say it." The next man to myself was the Padre—a fat jolly looking fellow of 30 years of age and 5 ft. 2 or 3 inches in height—who tucked his gown up around his waist and did his own part and helped me through with an air that showed him to be no novice at the business.

flavorings. Paraguay's oranges are so rich and sweet, said Washburn, no one ever suffers from hunger. "With oranges, a little yerba maté, a bit of *asado*, or broiled beef, or of boiled mandioca root, a Paraguayan would be perfectly contented; and if permitted to lie for three quarters of the time in his hammock and thrum his guitar, he would regard himself at the height of earthly bliss." (Washburn, i 434.)

*Ammen defined the curandero as a "curer," a native purveyor of plant and animal folk remedies. One practitioner he interviewed had a book of remedies; they were, said Ammen, cures in vogue in England 200 years before. (Ammen, 267.)

†Yerbale (from *yerba* or *hierba*, grass, weed, herb) describes an herb farm, generally, but in the La Plata area it refers to maté, Paraguayan tea. The plant is a South American holly whose leaves and shoots, when steeped, have aromatic and stimulant properties (like tea and coffee). Maté is the Quechua word for "gourd."

In Paraguay maté is universally drunk between meals: a morning drink, *cocida*, boiling water poured on leaves, is served with milk and sugar; *terraré* is a cool summertime drink; *maté amargo*, "bitter maté," is the unsweetened tea served hot or cold. See also n., p. 72.

The next day, Sunday, Powell added, "Horse racing has been going on all day in the plaza." It was a repetition of Caraguatay, and seems to have been a favorite pastime with the people.

On July 1 the travelers set out again, accompanied by a few officials and the priest, Daniel Sosa, who rode with them a few miles to put them on the right trail before saying farewell. Later that day they stopped at the house of Manuel Antonio Vera, who had a reputation as a medical man. Powell was hoping to obtain more medicinal herbs.

The only novelty they saw was a simple mill* for pressing sugar cane, "the first thing in the way of machinery I have seen since leaving the Capital—and it is of most simple construction," Powell noted.[28] Señor Vera gave them a guide to assist them through the Caraguatay Mountains. They crossed more of the ever-present streams and finally arrived at San Joaquín, where a letter of introduction to the local authority obtained for them a comfortable house for the night. Padre Sosa arrived the next day and accompanied them across the thirty miles of plains and streams (the Ihú River was so winding they crossed it twice) to the town of Ihú, where the official, Señor Alonzo Benitas, gave them a kind reception.[29] The men found this same kindness continually, and it was one of the things that left a lasting impression with them.

The following day a party of seventeen, headed by Señor Benitas, went into the mountains to hunt deer and jaguar, "the lords of the country." Several times the horses started and often fresh tracks were seen, but the hunters returned empty-handed. On July 9 the expedition moved off to the southwest across "fine rolling country" abundant in the deer they were seeking the day before, and so full of wild fowl that their dog "stood a partridge at every fifty yards." Several hours of riding brought the travelers to Caaguazú, a tiny village of seven houses and fifty people. Most of them had been sent there by the government, "one of the modes of punishment being to order a man and sometimes a whole family into an unsettled portion of the country, in order to commence its settlement." A man named Juan Cardozo took the men into his home.[30]

Two days later the party was on its way again, crossing the "perennial streams" which tumbled down the mountains they were negotiating. Powell found in this region a number of beautiful butterflies which he added to the collection. Another day's journey brought the riders to Villarrica and to the home of Don Luis Homen, who had so graciously received members of the expedition in February. Here Powell learned that the *Pilcomayo* had returned, so he sent a "runner" to Asunción to find out from Page if he should continue his explorations southward.[31]

It was not until ten days later, on July 21, that the answer came. The *Pilcomayo* was working badly and had to be repaired, so Page detailed a few men to this task and left on the *Water Witch* for Corrientes on July 14. There was, consequently, no reason to return to Asunción, so Powell decided to go on. He left the next day for Yacaguazú, due south. In this village he obtained a guide, an ex-army sergeant, who was with them for five days until they reached the town of Yuty. He showed them across swollen creeks and through mountain passes, and except for one evening when they camped on the banks of the Pirapa, he found them some shelter for the nights. When they reached Yuty the town official, Diego Martínez, gave them a room in the school.[32] After the observations were finished the group moved on, crossing the Tebicuary River, where they shot a toucan, and stopping at the home of Señor Luís Martínez (probably a relative of the Yuty official). This man's grand-daughter was married to a Brazilian named Urbano José de Moura who told the men much about the surrounding country. Rain kept the travelers at the house for another day, and on the last day of July they set out with Señor Moura taking them to San Pedro del Paraná, a village of ten houses and fifty people.[33]

Traveling mile after mile over open country, as they were, with only a simple isolated

*The Spanish *trapiche* (horse mill) or sugar mill involved two or three animal-powered vertical rollers, between which the cane was crushed. The juice was collected in a trough, boiled, and poured into hemispherical molds (see n., p. 30).

dwelling or a single government post house,* or a small village like San Pedro to intrude into the emptiness now and then, the men must have found the countryside of Paraguay like an almost deserted land. This, in turn, must have been a puzzling thing, considering the richness of the land in resources and natural beauty. But then, it was only fourteen years after the death of the man who had sealed off the nation from the rest of the world. As Page remarked, "Francia's policies threw the people on their own . . . but no skill or ingenuity was awakened."[34] The travelers were soon to see the remains of a Paraguay that might have been, had its history developed differently.

After a day and a half in San Pedro, where a crane and a hawk were added to the specimens, the expedition headed southeast toward the territory which was once the center of Paraguayan culture. It was here the Jesuits built their *reducciones*, reduced to ruins by time and desertion because of Brazilian slave raids and the threat of alien armies.† López at one point ordered Encarnación evacuated, mainly for military reasons.‡ "He who dares look back to Encarnación will be turned into one of the pillars of the state—a soldier," Page cleverly remarked.[35]

Through refreshing forests, across expanses of marsh, and over high rolling country, crossing the "perennial" streams like the Yoyta, the Cahi, and the Taquary, the riders jogged onward until on August 3 they reached the town of Carmen, really a city with its 1000 inhabitants.[36] Here they saw some of the fine old structures which had once caused Page to say, "One is lost in wonder and admiration at the will, the nerve, the zeal to attempt and the intellectual skill to achieve such master works."[37]

Two days later, after crossing the Taquary again and the Aguapey, the travelers came to San Cosme, where Captain Iranias welcomed them and gave them lodging in the school. This building and those attached to it were "regular and spacious, being constructed of fine red sand-stone, abundant in this section," presenting "a better appearance than most of the buildings of this country." The Paraná River ran close by, and the town was "prettily situated so as to overlook

*Post houses furnished food and lodging and fresh horses for couriers and travelers; their plan was fairly regular. Each installation consisted of two long thatched "cottages" about a foot apart, their common sides open to each other. One housed a kitchen and lumber shed, the other, the attendant's family bedroom and "living" area, strewn with hammocks and, in the rafters, lassos, bolas, and horse gear. Outside, next to each cottage, was a 10-foot high platform, the *andamiada* ("scaffold"), for sleeping when mosquitoes were unbearable. (Mansfield, 321-22.)

†The Jesuits (Society of Jesus, founded 1500) had been at work 1554 in Brazil. In 1608 they began methodically to convert Paraguay's Guaranís, and had succeeded in christianizing and settling many in communal villages (called reducciones) along the Paraná, Paraguay, and Uruguay rivers. Because of this monopoly of native manpower and productivity, however—and their extraordinary efficiency and ingenuity, the Jesuits as early as 1721 had irritated Asunción colonists who were effectively prohibited from profiting in Indian trade and slavery. In this officially recognized "Christian Republic" within Paraguay approximately 100,000 Indians were living in some 30 communities, each governed—albeit absolutely—by only two priests.

In 1750 Spain agreed to transfer seven missions on the Uruguay River to Portuguese control. Because the Guaranís here had long been raided by Brazilian "Mamelukes" (slave traders), they and their Jesuit guardians fought the transfer determinedly. To defend themselves against slave raids the Indians had been formed into companies and (by special dispensation) given firearms—all organized, outfitted, and drilled by the Jesuits! To carry out the transfer a combined Spanish-Portuguese army effectively destroyed and scattered this resistance (the Guaraní War, 1754-57), but in a new treaty (1761) the transfer was cancelled. This incident—along with the Crown's growing interest in the vast wealth reportedly accumulated by Jesuits in the Americas—led to the complete expulsion of the order in 1767. Much of its work was taken over by the Franciscans.

The country today has strangely reverted to former times: Paraguay has only one priest for every 6700 people.

(Weil et al., 111-12, 116; Wilcocke, 273-79; Caraman.)

‡Nuestra Señora de la Encarnación ("Our Lady of the Incarnation") was founded 1614 by the Spaniards opposite Itapúa, one of the 30 Guaraní reducciones on the Argentine bank of the Alto Paraná. In a treaty López agreed to establish a free port on the Alta Paraná—with Encarnación his only city on its banks. When he changed his mind he designated a new town, Carmen, to be the "port," and—at least temporarily—"abandoned" Encarnación.

It appears the ***Water Witch*** did not ascend this branch of the Paraná where, after about 280 miles, the spectacular Yguazú Falls dramatically marks the point today where Paraguay, Argentina, and Brazil meet. (Page, 222.)

this river and had not the overthrow of the Jesuits arrested its construction it would undoubtedly have been the finest town of the missions."[38] The church, three hundred feet by seventy feet, was especially attractive, and its façade was ornamented by columns twenty feet high and two feet in diameter, "of one piece and beautifully cut."*[39]

On August 7 Powell left San Cosme and traveled over the undulating country along the Paraná to the mission of San Rafael, which had remains of gilding, statuary, and wood carvings.†[40] Many of the missions had retained "in remarkable preservation evidences of the wonderful zeal, perseverence, skill and ability of their founders, in the structure, carving and painting of their churches."[41] They were a silent tribute to the early Jesuits and proved that "in art those missionaries were the Medici of La Plata."‡[42]

From San Rafael the party turned north and in the succeeding week passed through Santiago, Santa Rosa, Santa María, and San Miguel, stopping at several estancias overnight. Geographical observations were made at all of these places. On August 15 they arrived at an iron works.[43] The superintendent of the shop was a Swede named August Liliedat. The foundry was not in operation at the time, but it had already cast "some thousands of cannon balls."§ The workers were cleaning out a nearby creek to make it navigable for large flat-bottomed boats to transport ore and finished products. Powell obtained a specimen of the ore for the mineral collection.‖[44]

*Francia had many Jesuit churches torn down. San Cosme, near Encarnación, contains the only structure in Paraguay still possessing its original roof. It attracts many visitors today. (Weil et al., ix; Kolinski; Warren, 187; Page, 225.)

†The Jesuits had shown their charges all the Western manual arts and trades, finding them expert craftsmen and workers capable—to this day—of meticulous imitation. Aside from Spanish (see n., p. 48), however, the priests failed to teach the most important survival skill of all: self-determination in a modern society. This profound observation would have been quite academic had not this failure produced exactly the resulting evil—generations of dictatorships—the Paraguayan people were obliged to suffer.

The word "guaraní" meant warrior, and both Indians and Spaniards—whether pure-blood or mixed—had duly exercised their rights in colonial times. "This seems evident," said Washburn, "from the insubordination to authority that showed itself whenever power was grossly abused; and it was only after the influence of the Jesuits had emasculated the general mind of all sense of responsibility and every feeling of personal reliance, that the whole race became the willing forgers of their own fetters. . . ."

(Hutchinson, 14; Washburn, i 66; Page, 505; Weil et al., 116; Caraman.)

‡The Italian Medici family, unscrupulously wealthy and powerful throughout the 14th to 16th centuries, ostentatiously patronized European intellectual life, literature, and the fine arts. Page here—when he saw a dozen richly carved high-backed chairs—seems to have admired a similarly devious encouragement of the arts, but he added, "It may be noticed at this point that intrigue and cunning are words familiarly and unhesitatingly associated with the Jesuit; but nothing is hazarded in saying that in their labors among the La Plata savages an energy, piety, zeal, and perseverence worthy of the best cause are eminently conspicuous."

Washburn found Page's praise characteristic of Southerners who, like the Jesuits, he said, believed "the highest civilization . . . is that where the great mass of the people are slaves, and the few live in luxury and idleness on the bread they have never earned." Nevertheless, events here—eight decades after the Jesuit expulsion—prompted several expedition participants to disparage both Jesuit aims and results.

(Page, 120, 466, 472-73; Washburn, i 130; Kolinski.)

§Despite considerable iron ore, Paraguay's colonists had made little iron, causing the imported product to be extremely expensive. In 1849 an Englishman, Henry Godwin, was engaged to start a foundry at Ybicuy (Spanish Ybycuí). Although he fully expected to train his men, insurmountable labor and materials shortages defeated him after two years. The government next employed Liliedat in 1852 and in two years he was producing ingots, cannon balls, and small cannon. His workforce, however, was almost entirely convict labor, and with an indifferent (European) staff he too quit, in 1854. Liliedat had also been an avid naturalist; he studied the medicinal properties of plants (practicing on his workers!) and gave some specimens to Page.

Anxious for a domestic source of weapons and munitions, President López had instructed his son in Europe to recruit the best technicians available—British mechanics and engineers. They soon ran the Paraguayan foundry, shipyard, arsenal, railway, medical facilities, architectural operations, and military establishment. Hopkins' fear that England would "invade" his U.S. marketing domain (n., p. 7) was well founded.

After 1870, however, the war-torn foundry was abandoned.

(Plá, 11-20; Page, 236-37.)

‖In modern times Paraguayan iron ore has been judged of high quality but it is still largely unexploited.

The travelers turned northwest for the final phase of the journey. Most of it was over broad sunlit plains dotted by the towns of Ybicuy, Baehu, Paraguarí, Piraju, and Ytagua. It had been a long trip, and even the horses were "sensibly fagging." The group had been en route two months and twenty days, and had covered 720 miles. They arrived in Asunción at 4:10 p.m. on August 21.[45] The *Water Witch* was still gone, and a critical situation which was to affect the whole expedition had evolved.

After the *Pilcomayo* returned from her Bermejo River trip she was employed for the next few days in transporting wood to the Chaco side of the river. There had been, it seems, some question of theft, and to avoid any further trouble the supply was moved. It was evident, however, that the small steamer was not working well at all, and Page detailed engineer Stump and five others to overhaul and make alterations on the vessel. President López offered to give any assistance necessary to repair the *Pilcomayo*.[46] With this assurance that things would be attended to, Page detached a few officers to copy the charts of the work done up to then and left on the *Water Witch* on July 14, arriving at Corrientes two days later.[47] His purpose was to explore the province of Corrientes, and possibly that of Entre Ríos. For the first two weeks the weather was bad and Page himself was not feeling well. The Bermejo trip must have affected several men; some were sick shortly after, and Acting Master George Welsh was ordered back home because of poor health.[48] To Page the time was not lost, however. The expedition had collected a large quantity of specimens, and Page prepared lists to send along with them to the United States.

The specimens were grouped into five boxes: birds, reptiles, seeds and medicinal plants, flowers and other plants, and geological specimens. In his long accompanying letter Page made no comments about the birds and reptiles, but he mentioned several things about the other boxes. He was sending yerba maté seeds, "but they tell me they must be planted within ten days." He had wanted to send tobacco and cotton seeds which had been promised to him, "but there is but little confidence to be placed in the promises and assurances of these people. Everything is postponed until tomorrow; and very often that never arrives."

The medicinal plants sent were used as cures for almost every ill imaginable: purifying blood, tetanus, nervous sedation, cough, ulcers, indigestion, rheumatism, sore throat, diarrhoea, and on and on. Then there was a list of flowers, 134 different varieties from Paraguay and 66 from other places, with the color, description, and provenance for each. In addition there were plants collected in Paraguay and along the Paraná and Bermejo rivers, including especially the *Victoria regia*. Finally, there were 18 varieties of beans and peas. Almost all the botanical names were in Guaraní.* The geological specimens had come from Coimbra, Corumbá, Pan de Azúcar, Encarnación, the iron works, and the banks of the Bermejo. Soil samples were being sent to Maury, who had "great interest in analyzing earths."† It was a monumental job of cataloging—no wonder it took Page two weeks to do it![49]

*Guaraní ironically was originally an oral language. The Jesuits and Franciscans had reduced it to writing, furnishing grammars and dictionaries. They translated everything into Guaraní and natives in their care never even spoke Spanish. (Priests claimed the Guaranís incapable of learning Spanish—but they had no trouble teaching them a great deal of church Latin!) Today, dependence on the native language is still strong: fully half the population is bilingual. Spanish is spoken only in urban society and for official and commercial purposes. (Page, 503; Weil et al., 10, 30.)

†Maury (see n., p. 9) was at work charting ocean winds and currents, keeping a dozen lieutenants combing ships' logs for data. As a result of this work he brought out the first oceanography text in 1855; his analysis reduced the New York-San Francisco route (around Cape Horn) from 180 to 133 days. He also mapped the floor of the Atlantic Ocean, revealing a route for the 1866 submarine cable. His Pacific Ocean charts proved extremely useful in World War II! Samples of inland soil do not seem pertinent to his work, but see n., p. 86.

(Maury, "Annual Address," 16; Ponko, 228.)

By August 3 both Page and the weather were fine, and he took a group of men and set out downriver in a long boat. The exact route of this expedition is not clear, but at some point, possibly Bella Vista, he left the water and traveled inland. He was back in Corrientes by August 13, when he wrote to Dobbin to say he had been through a region of the province which was largely marsh and swamp land, sparse in population. With few laborers production was obviously limited, and exports were mainly hides, tallow, and lumber. Much of the land, he thought, would be good for tobacco, cotton, and corn, and he felt that European immigration might partially solve the manpower problem.* In one place, however, where a Frenchman was using a new type of plow his native neighbors thought he was crazy.[50] On a few estancias he found orange groves, but they were much neglected. He had pointed out once before—and he emphasized again—the fact that the recent civil wars had much to do with the depopulation and desolation of the land.[51]

On the same day that Page was writing to Dobbin a special messenger arrived from Asunción with a letter from Hopkins. He told Page that "the entente cordiale between the Paraguayan Government and myself has unfortunately come to an end."†[52] Page felt it was merely a matter of misunderstanding between Hopkins and López. He saw no reason to side with the consul against the President. "Thus far, to myself and the Expedition the President's course has been all that I could have asked. He has voluntarily offered to do all that lay in his power, to promote my views."[53] Page dismissed the matter and started on another inland expedition.

This time Page's travels took him through northern Corrientes. The journey was facilitated by the fact that there were post houses‡ every two or three leagues with extra horses.[54] Page found that Argentina, too, had its sunlit plains and perennial streams. The region was picturesquely dotted with small, shallow lagoons filled with pink *Victoria regia.* The lagoons were linked by a chain of narrow creeks.[55] A large lake called Iberá§ interested Page particularly because a veil of mystery hung around the region and no one had gone near it for a long time. It was reported that religious rites had been performed by the Jesuits on an island in it and that it was somehow

*A large party of French émigrés recruited by Solano López had come to Paraguay in 1855; they were badly treated. The English technicians fared just a little better. But not until after 1870, when Paraguay was exhausted and destitute—having lost one-half her population (mostly young men) and one million pounds sterling (in notorious Baring Brothers loans)—was the government driven to sell large numbers of tracts to foreigners. For Europeans contemplating settlement Washburn was unusually sanguine: "The nation, with its peculiar civilization, is destroyed. The Paraguayan people no longer exist in numbers sufficient to control the destinies of the country. . . . [It] offers inducements to the emigrants from the Old World greater than any other part of South America, if not of the world. . . . Paraguay may soon become the paradise of South America."

Few Europeans came, however, until the twentieth century. Since then Australians, Germans, Austrians, Japanese, Italians, and others in large numbers have made Paraguay their home.

(Plá, xxiii; Kolinski; Washburn, ii 605-606.)

†Hopkins' abrasive character had undoubtedly aggravated his difficulties with the Paraguayan government. His own employees had reported his "swaggering, bullying way," his "tyrannical and overbearing" conduct. But López, said Peterson, "may have feared that the Americans were bringing his people ideas of liberty and natural rights as well as the skills they so much needed." Ynsfran thought this unlikely, as López had not similarly harassed the British technicians. They, however, were held in check one by one with strict individual work contracts (and miserable pay—half in cheap paper currency), whereas Hopkins represented a larger, organized threat with both "yankee ingenuity" and strong support from home. The President deliberately hampered the company's operations and thereby destroyed the enterprise. The United States and Paraguayan Navigation Company promptly filed claims with the U.S. Department of State for losses incurred and for profits it might have made. (Washburn, i 369; Peterson, 257-58; Ynsfran, 250.)

‡Accommodations at one post house (see n., p. 46) were quite informal: "[We were] refreshed with supper of 'asado' and mandioca, and, after cigars, made quite members of the family by having our hammocks slung up in the same room with the master of the post, his wife, three other women, and five children; one of these an infant, who entertained us during the night with solos and snatches of song that indicated good lungs." (Page, 267.)

§Iberá or Yberá is Guaraní, *y*, water, *bera*, brilliant; the lake, however, was so overgrown with aquatic plants it was more like a marsh. (Page, 300.)

sacred. "The Jesuits left a mysterious religious aire hanging around it, which no man has dared breathe through," he wrote to Maury.[56] About forty miles east of Corrientes Page visited Don Angel Bedoya's estancia Yrisbuqua, one of the largest in the province.[57]

About this time another letter and copies of the newspaper *Seminario* from Hopkins reached Page. Hopkins was being attacked by the Paraguayan press and wanted Page to support his arguments—with the guns of the *Water Witch* if necessary. Page wrote to Dobbin:[58]

> If Mr. Hopkins expects to involve me and the Water Witch in the disgraceful affair between the Government of Paraguay and himself, he deludes himself with very false hopes. . . . By a most impolitic and unauthorized course, he has brought upon himself the rath [sic] and indignation of a Government, which has the power—because of its peculiar relations with its people—to embarrass and render profitless the entire enterprize [sic] of those American citizens, who have so unwisely put him at its head.

Page continued his journey to Itatí on the banks of the Paraná, and while he was here a messenger arrived from the *Water Witch* with documents and letters which made him realize he had to return to Asunción. Once again the elements seemed to presage the outcome of events. Rain, thunder, and lightning accompanied the steamer upriver. With a full salute, she entered Asunción September 20.[59]

Pan de Azucar.

7. TROUBLE IN ASUNCION

By the time Page got to Asunción the trouble that had been brewing for two months had already come to a boiling point, and though he had wanted to stay out of the whole affair he found himself forcibly very much involved.

It all began on the afternoon of July 22 when Clement Hopkins was accompanying Madame Guillemot, wife of the French Chargé d'Affaires in Asunción, from the American company's headquarters in San Antonio to her home. As they galloped along they encountered a herd of oxen which were startled and scattered in several directions. Augustín Silvero, a soldier who was driving them, became angry and struck Hopkins several times across the shoulders with the side of his sword. Clement immediately reported the affair to his brother who wrote an indignant letter to President López demanding public satisfaction and Silvero's punishment.[1]

On August 5, José Falcón, the acting Minister of Foreign Affairs, sent Hopkins a reply telling him that Silvero would be dealt with properly and he should not concern himself about it any further.* But this occupied only a small part of the letter. A grievance that the government had been holding for many months came to light. Falcón asked Hopkins why he had made his statements in Buenos Aires about the United States refusing to ratify the navigation treaty, telling him that he was forcing Paraguay to change its mind about the matter.[2]

Two days later a proposal was sent to Hopkins that he sell his rights in the United States and Paraguay Navigation Company to the Paraguayan government.[3] Hopkins refused to reply, but he wrote to Marcy that the reference to the treaty was "a jesuitical pretext, to continue and cement the system of Doctor Francia, in making the State the only merchant," and that he was waiting

*López later told Page the soldier had received 300 "stripes," running the gauntlet of his regiment. (Page, 271.)

for "a farther development of the violence of President Lopes [sic], as directed first against other American citizens, and lastly against myself, for taking up their defense."[4] He wrote at the same time to Page, describing the situation.

Hopkins did not have to wait long for the "farther development" he was expecting. On August 16 López declared the sale of the land to Hopkins and his company annulled and demanded that it be returned to the seller. He enforced his declaration by sending troops to occupy the premises at San Antonio.[5] Hopkins wrote a hasty note to Falcón reminding him that treaty or no treaty the United States would maintain an interest in its citizens and would give them its "powerful and constant protection."[6] He then sent his second letter to Page who, he felt, represented this "protection." Page, however, refused to act because—and it is to his credit—he saw the basic issue very clearly: namely, that it was Hopkins the businessman who was involved in the affair, not Hopkins the representative of the United States.

Things were going badly, and Hopkins suddenly assumed a martyr complex and wrote a long missile to Marcy telling him the whole story of how he had established the company which was operating so successfully and of which López was now jealous, of how Americans had been attacked at night on the street and had things stolen from them with no attempt at justice, and finally, how he himself had been personally affronted by both López and his own countrymen. In the newspaper López had called Hopkins a liar and said he had engaged in smuggling under his official passport.

Hopkins sent copies of the paper to show that two separate editions had been printed: one for the general public and one for foreigners, especially the Americans. Most of all, he resented the fact that his official position was completely ignored—particularly by Page's men. On his one visit to the *Water Witch* while Page was away no honor or formality of any kind had been accorded him, "neither side boys* nor the boatswain's pipe"; no officers received him, nor was the commander in uniform. Yet when the Brazilian consul visited the ship he was given a nine-gun salute. Nor did the men ever greet him in the street, in spite of the fact that he let them make use of his home. How could he ever be expected to put up with all of this? He concluded his letter by resigning.[7]

López' personal antipathy for Hopkins had now reached a climactic point. On August 24 a decree demanding that all foreign establishments which had been guilty of "quiet and notorious operation" for many months obtain a permit or license to continue in existence was proclaimed in old colonial style, by a trumpeter and posted "like a handbill at three corners of streets in this city."[8] Two more decrees were issued within the next week, and then the ax fell. On September 1 Hopkins' *exequatur* was revoked. In the eyes of Paraguay (López) he no longer held a diplomatic position.[9] José Falcón sent a letter to Marcy telling him that the action taken against Hopkins was of a personal nature and that Paraguay still wanted to maintain friendly relations with the United States. It was, he said, in the interests of peace and good relations that the exequatur had been terminated.[10]

When Page received this news from Hopkins he realized that the Americans in Asunción might actually need his help. Then, too, since López had shown himself favorable to the expedition all along (barring his annoyance at their entering Brazil), Page thought he might be able to do something about the situation. Powell, who had returned from the interior just a few days before the decrees, offered to mediate for Hopkins, but he refused.[11]

Diplomatically discredited, Hopkins had to face further difficulties when, on September 11, López demanded an inventory of the company supplies.[12] The day after he arrived in Asunción

*Two or more crew members assigned to stand at the gangway as a gesture of respect. As a former midshipman Hopkins would have relished this honor!

Page visited López, who told him that Hopkins would have to cease operations and leave the country. The President assured Page that the other Americans in Paraguay were in no danger whatever and that he still viewed the expedition favorably. When Hopkins was told of the President's decision he said it would be impossible to leave, since no ship would risk the possibility of incurring López' displeasure by having assisted the departure. Page went once again to the President and asked for assurance of safety for anyone who would take Hopkins aboard. López replied that the ex-consul could leave any time he wished and would merely have to clear his belongings through customs. At last everything seemed arranged. But when Hopkins tried to clear his property he was told he would have to hand over the title to the land he had bought and any other papers relating to the company.[13] This was too much. Page became angry at this treatment of the company and its personnel and found himself forced to side with Hopkins.

Ironically, at this crucial moment Page received a note from William Trousdale, the United States Minister in Rio, telling him that the Brazilian government had given him permission to explore the waters above Albuquerque. While the news was welcome, it was also frustrating. Page would now need López' permission to reascend the Paraguay. Under the present circumstances this would not be easy. As he told Maury, ". . . just at the time of all others when I should be able to keep this old villain in the best humor, we have this infernal rupture with this man Hopkins."[14]

On September 28 Page wrote Falcón two letters of very different tenor. In the first he informed the Minister of the permission from Brazil and asked for authorization for the *Water Witch* to pass "through that portion of the territory of Paraguay which lies between Asunción and the territory of Brazil." In the second letter Page protested against the demands made of Hopkins and the company: "The papers are the private property of Mr. Hopkins, which is a sufficient reason that he should retain them; and as such cannot be surrendered to the Collector of Customs." He added that if the Americans (Hopkins included) were not allowed to leave with their things he would take them on the *Water Witch.*[15]

Later that day Falcón called Edward Bushell, Page's secretary, and asked him for translations of Page's letters. Page refused to let Bushell translate them "because he is not perfect in such things, and will soften their meaning so favorably–for the Government–such being his idea of avoiding difficulties–that the language will scarcely be the same."[16] The letters were returned unanswered. Page was more angry than ever. He argued that any country communicates in its own language, and he knew that Paraguay had translators. He considered Falcón's action as a "contempt of official communications unknown to the civilization of the age."[17] There was only one thing to do.

Page wrote Dobbin, "There is no other course for me to pursue than to receive the Americans, on board, and give them a safe passage out of the Country. Were the Water Witch not here, I do not believe that one of them would get out of it."[18] Late that same day the steamer "received for safe keeping 3 boxes and one black bag belonging to Mr. Hopkins the American Consul."[19]

The following day more personal belongings of members of the company were brought on board. Shells and shrapnel were moved into place near the forward and after guns.[20] If there was to be a showdown, the *Water Witch* would be prepared. This activity was undoubtedly reported to López and at midnight he held a consultation with his confidential advisers and resolved to give Hopkins a clearance for his things, even including 800 arrobas of fine tobacco.[21] Page considered this a real victory. "Such a 'come-down' has never before been known in Paraguay. . . . The foreigners came on board to see me, before leaving, and said it was the best thing that had happened for all foreigners."[22]

On September 30 the last of the personal possessions of the Americans were brought aboard and they themselves came onto the ship around noon. By order of López the streets were deserted. "Not a man, cart or horse was to be seen."*[23] At 2:00 p.m. the *Water Witch* weighed anchor and sailed away from Asunción, one day short of being a year that she had first arrived there.[24] When the vessel passed Tres Bocas two days later Page prepared against an attack, and in fact had determined that if the Paraguayans tried to interfere in any way with his plans he would capture their whole navy![25] This was not the first time the idea had occurred to him. Some days previous he had written to Dobbin, "We could take possession of what is called their Navy—and thus induce suitable terms." It was, after all, only five schooners and a brig.[26] No incident took place, however, and the steamer anchored at Corrientes in mid-afternoon of October 2.

Corrientes was now to become the expedition's headquarters. The day after the *Water Witch*'s arrival a house was procured to store various articles, and the *Pilcomayo*, which had been brought in tow with baggage and equipment, was cleared for continued repair work.[27] Page wrote several lengthy accounts of what had occurred and told Lieutenant Moore to take them personally to the United States. He also took along with him the little gift that Antonio Tachiera had given them in Concepción for the United States President.[28]

Any hopes that Page may have had for eventually returning to Brazil were shattered on October 3 when the Paraguayan government issued a decree excluding all foreign vessels of war from navigating the rivers of the Republic or exploring the waters of the upper Paraguay "whether they are in the territory of the Republic or in that of adjoining States, through the main river of the lower Paraguay."[29] While no names were mentioned, the decree was clearly directed at the *Water Witch*.† Page had previously written to Maury, "The President is more absolute than any Monarch on the Globe"; "this man has no more idea of a Republican Government than one of the Wild Indians of the Chaco."[30] To these sentiments he now added:[31]

> I would give my little finger if the Navy Department would only say to me—you ascend the Paraguay, so far as navigation will admit of, under the permissions granted you by Brazil. All Paraguay with her 6 men of war would not stop the Water Witch: and if they attempted it, they would lose the whole half dozen.

Page had wanted by all means to attempt an ascent of the Pilcomayo River since he thought it would open the whole interior of Bolivia to the outside world. The proclamation now made this impossible. It did not, however, put an end to the expedition by any means. Argentina was open to them, and Page had already told Dobbin that "the waters of the Uruguay and Salado rivers are streams of deep interest to Commerce and Geographical knowledge."[32] In addition, he still hoped to explore the upper Paraná and perhaps the rest of the Bermejo, though both these rivers presented problems because they were sore spots between Argentina and Paraguay. The Bermejo especially was a touchy question. "López made a Treaty with Urquiza . . . for both banks of the river Vermejo: and Urquiza—one can't judge how he could have been so blind—signed the Treaty: but when it came to be ratified by the congress they would not touch it." Nevertheless, "the Government of Paraguay has most arbitrarily seized upon this river, and now excludes all vessels from its navigation. . . ."[33] Had Page remained on good terms with López there would have been no problem, but now the situation was quite difficult.[34]

*Unsure of Page's reaction to these difficulties, López had actually thought the *Water Witch* might fire upon the city. He quickly got a mechanic to examine the running condition of his carriage; he was too heavy to ride a horse. (Page, 276.)

†Page's persistent desires to explore the upper Paraguay were in response to Maury's instructions (n., p. 11). López refused to grant such a precedent, fearing his enemy Brazil would ask the same privileges. (Washburn, i 365; Page, 118.)

> To all of which I am indebted to that miserable fellow sent out here first as consul, then as *General Agent*: as well fitted to one as the other—a disgrace to both. And that a fellow so contemptible should have done so much mischief[!] He involved Americans in a serious difficulty which forced upon me their protection.

Nor was this the only effect of the trouble in Asunción. In mid-October a man named Buckalew arrived in Corrientes with a letter for Page from the Department of State authorizing him to make the exchange of the ratifications of the treaty of March, 1853. Page at once wrote to Falcón informing him of the fact and asking if the *Water Witch* could come to Asunción for this purpose. He sent Murdaugh to deliver the letter personally.[35] Apparently Dr. Carter went along.*[36] They came back with a letter from Falcón saying he was returning Page's letter to him because it had not been translated and that he was astonished that Page would persevere in his determination to "mortify" him.[37]

Page wrote to Marcy telling him what had happened and explaining that not only did he see no reason to write in Spanish, but that he did not feel his secretary was capable of making an accurate translation. He suggested asking the Commodore of the Brazil squadron to come on board the *Water Witch* and steaming into Asunción to force an exchange, if necessary.[38] He wrote at the same time to Dobbin saying that Falcón's action was an insult to the United States, since Paraguay had long ago ratified the treaties made at the same time with England, France, and Sardinia. "There are governments whose respect can be secured, only, by an exhibition of a power adequate to the enforcement of respect: and Paraguay stands eminently in this position."[39] There was, in fact, soon to be a show of power, but not at all as Page expected.

The last echoes of Hopkins were heard in Rio. He wrote to Marcy saying that he had asked the opinion of people in Corrientes, Buenos Aires, Montevideo, and Rio about López, and all had agreed that the United States government should demand satisfaction for the persecution of a U.S. consul. The Brazilian government, he said, was "grieved" that Page had submitted to the Paraguayan decree. Then, in true Hopkins fashion, he suggested beginning negotiations with Brazil for the commercial navigation of the Amazon, assuring Marcy that they wanted American capital there.[40] It was almost as though he had started the old wheels spinning all over again.

*Page makes no mention of Dr. Carter, but the ship's log says he left for Asunción "in observance to an order from Capt. Page."

Attack upon the Water Witch.

8. GUNFIRE AT ITAPIRU

While the correspondence on the treaty was going on, the *Water Witch* went up the rivulet Ataja, five miles from the mouth, to cut wood for fuel. She was gone on this mission for ten days. After she returned it was another three weeks before things were sufficiently settled for her to leave for La Plata waters. At 7:30 on the morning of November 12, with Señor Catral, a member of the Argentine congress, as a passenger, the steamer sailed for Buenos Aires.[1] At La Paz the men discovered that the coal hulk was filled with water and had sunk to the bilge. They decided to leave it and continued on their way, stopping at Paraná to land Señor Catral. The inner roads of Buenos Aires were reached on the seventeenth. Once again there were desertions—three of them.[2]

On November 20 the ship moved to Montevideo for a complete overhauling and for supplies. In addition to the usual foodstuffs the steamer this time took on iron spoons, needles, thimbles, thread, combs, twine, blankets, and varnish, among other things. During the stay in Montevideo—almost a month—two more men deserted and three new crew men were taken on. Back in Buenos Aires the ship had to transfer two men to the *Germantown* because they were "condemned by a medical survey." One of these was John Bowditch, who had accompanied Powell on his long trip through Paraguay.[3] There was something of an exchange, though. An officer from the *Germantown* named William Jeffers was soon to join the *Water Witch* and guide her through what were perhaps the most exciting moments of the expedition.

Three days before Christmas the steamer started up the Paraná River. At Rosario she took on the representative of the province of Salta as a passenger to Paraná city, disembarking him there on the twenty-fourth. Christmas day was spent off La Paz where the ship remained for three days. During this time an effort was made to bail out the half-sunken coal hulk but finally it was abandoned. On New Year's eve, at 11:00 p.m., the wheels of the *Water Witch* once more came to a stop in Corrientes.[4] The next day, the first of the new year, William Jeffers, now raised to the rank of lieutenant (promotion from master was made officially on January 30),[5] reported for duty and was given charge of the *Water Witch* under Page.[6] This was the position Lieutenant Moore had held. With four of her former officers—Ammen, Donaldson, Moore, and Welsh—gone, Jeffers was a welcome addition to the crew.*

When the *Water Witch* left for La Plata several men stayed behind, some to work on the *Pilcomayo.* Among those who remained were Lieutenant Murdaugh and Page's son, John. While the *Water Witch* was away these two took a trip—possibly on Page's orders—through the province of Corrientes. They started south, making their first stop at Bella Vista, the town "hid among orange groves," where they met Mr. Hall again. On the way they had passed through the town of Saladas which, in spite of its good soil, had only five hundred inhabitants, and through San Roque, with only three hundred.[7] From Bella Vista the travelers went south again to Goya, passing through Santa Lucía, where Murdaugh saw "the best church I have seen in the province." Goya, a town of 7000, is fifty-four miles from Bella Vista.

When Murdaugh and John Page arrived the governor of the province was in town. It was here the two spent Christmas, attending on Christmas eve, as Murdaugh put it, the "Mass of the Cocks" (a literal translation of the term used for midnight Mass),† and on Christmas evening itself a special ball. Not long after, the Governor decided to take the visitors across part of the province himself. The departure was a gala affair. The Governor and his secretary, Murdaugh, and Page rode in a large carriage drawn by six horses. In attendance were four officers, thirty-six soldiers, three servants, and two clerks. Murdaugh recorded, "Off we dashed at half speed, trumpets sounding, sabres clashing, ladies waving their adieus from windows and balconies."[8] To the fourteen-year-old Page it must all have been very impressive.

Across open and beautiful grasslands the party rode almost to the other side of the province, stopping finally at Curuzú Cuatiá. At this point Murdaugh left young Page with the Governor‡ and traveled on alone toward Restauración (today Paso de los Libres). Midway on his journey he stopped at the estancia of Aimé Bonpland,§ the famous French naturalist whom Francia had held a prisoner in Paraguay for nine years even against the protests of European nations. In November 1853 Page had invited Bonpland to join the expedition on the upper Paraguay, but the

*William N. Jeffers (1824-1883) was appointed midshipman in 1840 and served in the Pacific and with the Brazil squadron. He had taught mathematics at the U.S. Naval Academy, producing several texts on navigation, gunnery, and ordnance. Before joining the *Water Witch* Jeffers had just surveyed (for a board headed by E. G. Squier) the Bay of Fonseca and Porto Cortez, on the Pacific and Atlantic oceans, respectively, as terminals for a possible trans-Honduranean railroad system. He had requested assignment to the expedition. (Page, 302.)

†*Mesa de las Gallas.* (Page, 293.)

‡This somewhat mysterious trip, to place young Page in the governor's care for a few weeks, occurred prior to the attack described in this chapter—hence without any outside warning of trouble (except Hopkins' predicament). But while planning to sail the forbidden Paraná, however, Page may have felt his son safer inland. The governor, Don Juan Pujol, gave the boy a horse and furnished some Spanish instruction, but more importantly must have sent him to President Urquiza's college at Concepción (see n., p. 87) for much of that time. Page later wrote favorably of the school visit without specifying when it occurred. (Page, 295, 416.)

§Aimé Jacques Alexandre Bonpland (1773-1858) had traveled with Alexander von Humboldt 1799-1804 through Mexico and the Andes of South America, collecting botanical specimens. Settling down as a professor in Buenos Aires (1818-21), he established a ranch in the Candelaria area of Misiones (not part of Paraguay) to study the yerba maté. After two years Francia suddenly sent 400 men to Bonpland's colony, killing all the male workers (Bonpland himself was wounded), burning the buildings, and taking him in chains to Santa María, where he was freed but forbidden to leave the country. Bonpland built another small farm but in 1831 his property was again seized. He managed to return to Buenos Aires in 1832—after nine years' detention. Francia had claimed Bonpland was growing yerba maté commercially. (Robertson, ii 78-91; Washburn.)

eighty-two-year old scientist was in Brazil at the time and received the invitation too late to accept it. Murdaugh stopped at his home with the intention of inviting him to participate in an exploration of the upper Paraná. Bonpland told him that if it could be arranged he would be delighted to go along, and would furnish a guide.[9] The expedition, however, was stopped before it had scarcely begun.

Murdaugh crossed the Uruguay River at Restauración and went into Uruguayana, Brazil, which lies directly opposite the Argentine city. Both cities are built on banks sixty feet high which border the mile-wide river. Restauración had five hundred people and Uruguayana 1500.[10] Murdaugh was now about two hundred and forty miles from Corrientes, and as it was already early January he decided to start for "home." When he got back he had covered approximately five hundred miles.

As it turned out, there was no need for Murdaugh to hurry back. The month of January 1855 held little more than routine activities. Page was waiting for the completion of work on the *Pilcomayo*, which he intended to use to explore the Salado River running into the Paraná near Santa Fé. It was known that the stream came from a long distance, but no one had ever attempted to navigate it in any kind of large craft. While he waited Page used his time to prepare two more boxes of specimens to be sent to the United States. One box contained botanical specimens of seeds, shrubs, flowers, and trees taken from Paraguay and Corrientes. There were a hundred and two varieties, including such things as water lilies, juniper, lupine,* azucena, pitcher plant, paradise apple, ferns, grass, vines, and creepers, and dragon's blood. The second box had fifty-seven kinds of birds, many of them unusual, like the *pirybita* with its yellow eyes and green legs. There were birds of green, yellow, red, and hazel eyes, and yellow, red, and blue legs. There was a three-foot *guaycamayo*† and the unusual *cuaraimibu* ("flute of the sun").[11]

A moment of what may be termed cool cordiality occurred on January 20 when the Paraguayan ship *Tacuarí*, recently built in Europe, came into Corrientes. Francisco Solano López‡ was aboard with the woman who was greatly to influence his life and the destiny of Paraguay—Eliza Lynch. There was an exchange of formal visits, brief and correct.[12] Yet, perhaps, it was a sign that Paraguay had nothing against Page personally. But this was to change drastically within the next two weeks.

The following day Page, Powell, Murdaugh, Henry, Stump, Dr. Carter, and nineteen men left on the *Pilcomayo* for Santa Fé and the Salado River.[13] Even though the stream was a common boundary between Paraguay and Argentina, Page felt the authorization from Urquiza and the governor of Corrientes gave him a right to explore its waters. So the expedition departed.

*Lupine or lupin seed is used as a food; a blue pigment is also derived from it. Azucena is a white lily. "Pitcher plant" is used for several plants with pitcher-shaped leaves which trap and digest insects. The paradise apple is a small wild variety. Dragon's blood is a dark red substance taken from certain palm trees; the powder was used as a pigment and (in the photoengraving industry) to control etching.

†The guaycamayo or guacamayo is probably the macaw.

‡That is, he was returning from France ready to help his father control the government. In Paris he had met Elizabeth Lynch, a compelling and spirited Irish woman who would disgust the López family and shock Paraguay's simple society as well. "Madame Lynch," as she was called, has been portrayed as "a beautiful but ambitious wench," responsible for greedily encouraging Solano López' megalomania; see n., p. 76.

The *Water Witch* in 1853 had dramatically called steam navigation to President López' attention—just as Hopkins knew it would. As Paraguay possessed no navy, Solano López in England commissioned the *Tacuarí*, a fast steamer with shallow draft. It was a merchant ship, but with boilers below the water line it could become a fighting vessel, when needed. In 1858—as the spectacular U.S. fleet approached Paraguay—the *Tacuarí* collided with a British ship and López once again refused to discuss damages; a year later British warships blockaded the *Tecuarí* (in port at Buenos Aires), forcing the President to concede.

(Washburn, i 473; Plá, vi, xvi-xviii.)

There were others, however, who did not feel as Page did. He had made no secret of his intention to explore the Paraná, but President López was determined that he would not. He sent a note to the Commander-in-Chief of the national fleet warning him that the *Water Witch* intended to enter Paraguayan waters, and telling him that he should prevent it at all costs. He was not to let the Americans sleep or rest or even set foot on any beach of Paraguayan soil. The commander-in-chief *ad interim*, Wenceslao Robles, now at the general encampment of Paso de la Patria on the Paraná, was also to be warned.[14]

At 6:40 on the morning of February 1, 1855, the *Water Witch* weighed anchor and started up the Paraná. It was "cloudy but pleasant," and as the ship moved along soundings were made in order to check and rectify, if necessary, the earlier charts.[15] At 11:00 a.m. the steamer reached the confluence of the Paraguay and Paraná rivers and turned right into the Paraná branch. About four miles up the river the large island of Caraya separates it into two channels. As the *Water Witch* steamed up the right side, between the island and the Corrientes shore, the men on board saw across the river the semicircular brick fort of Itapirú with its six guns *en barbette*. Situated about thirty feet above the river, it had good command of the waterway.

In the fort itself Commander Vicente Duarte watched the steamer proceed up the river; from the Paraguay side a canoe slipped out of a small creek and the paddlers also observed the vessel's movement. At 11:30 the steamer ran aground on a sand bank. The crew spent about an hour laying out a kedge* and trying to get the ship free. They finally stopped for lunch. During this time they saw preparations being made in the fort that seemed hostile. Jeffers decided, as a precaution, to move the starboard gun to the forward port and ordered forty shrapnel† and twelve regular shells prepared. He had twenty-eight men on board, including five cooks and stewards; two men were sick. Potts was put in charge of the bell and on deck to assist the pilot. Samuel Chaney was at the helm, Lamdin took charge of the guns, and Taylor fired the engines.

At 1:20, free from the sand bar, the ship went back to the river separation and started up the other channel. As she progressed steadily a Paraguayan canoe came alongside and a man offered a printed paper in Spanish. It was a copy of the October 3 decree on navigation. Jeffers declined to receive it on the grounds that he could not read it. Robles later reported that the commander of the *Water Witch* "took the decree and threw it contemptuously at the officer," saying he was going to ascend the river anyway. The ship was then hailed by a man mounted on the fort's parapet, by this time about three hundred yards away. Whatever was shouted was not understood on the steamer. Again, according to Robles, the Paraguayans attempted to halt the vessel by firing from the fort three times, only to have this gunfire scoffed and laughed at by the men aboard the steamer.

Three times, Robles claimed, the *Water Witch* was ordered by voice to anchor. When she refused, the commander ordered a ball shot "to pass in front of the bow without touching the hull." The intervening shots and voice commands are not mentioned by Jeffers. When he did not stop at the first salvo two blank cartridges were fired in quick succession, followed by a *shot.*‡ The pilot wheel was destroyed, the ropes were cut, and Samuel Chaney was mortally wounded.

*An anchor dropped at some distance, so that taking in the line pulls the ship toward it.

†Naval gunners used shot (solid iron balls to punch a ship's hull), grape (clusters of smaller balls which scattered upon firing to cut sails and rigging), and shrapnel, which had just been introduced 1850 into the U.S. Navy. The invention of Henry Shrapnel, a British general, shrapnel was a hollow spherical "bomb" loaded with gunpowder and marble-size balls. Its fuse was lit by the firing of the gun and timed to burst over enemy heads.

‡"At what part of the vessel it was aimed," Page wrote with some sarcasm, "I can only judge from President López's dispatch to our State Department, where he magnanimously says it was directed so as to 'pass ahead.' If so, it struck wide of the mark. . . ." The piece had been directed by a former French naval gunner in Paraguayan service. (Page, 307.)

The pilot became frightened and hid, but Potts dragged him out and forced him into action. By this time the fort had been passed, but the water level was dropping and there was only one foot more depth than draft. The pilot told Jeffers, "We shall certainly ground."

Jeffers did not want to risk running ahead and grounding, because the fort was reported to have 6000 men, and he feared he might possibly lose the ship. His only course was to go back downstream. When he did so the fort's guns opened fire once more. The steamer now returned the fire, Jeffers himself directing the guns from the bow until he had to leave to force the pilot back to his station. The fort fired twelve guns. The firing was slow but "remarkably well directed." Two boats the *Water Witch* was carrying fell into the water in pieces, and one of the paddle wheels was disabled. The firing from the steamer dismounted one of the fort guns, and Jeffers thought that the shrapnel must have killed a few of the men inside. Robles, however, mentioned no casualties in his report.

Once she was out of range the steamer stopped to make some immediate repairs. Jeffers considered attacking the Guardia Cerrito at the mouth of the Paraguay, but the newly arrived war vessel *Tacuarí* was there, firing up her engines and readying her five guns. Jeffers felt that in her present condition the *Water Witch* was no match for the Paraguayan ship, so he abandoned his plan and prepared to leave as quickly as possible. At 3:00 p.m. the steamer started the return trip to Corrientes. A half hour later Samuel Chaney died from the effects of the gunfire. His body was taken ashore at Corrientes by Lamdin and Potts and arrangements were made for burial in the local cemetery.

Robles immediately sent a report to both President López and his son, giving his account of the "first essay of our land artillery," and commending the "valor and enthusiasm of the commandant of Itapirú and of the troops under his command," adding, "Such has been the retreat of the Water Witch, taking away a lesson which it had been sometime seeking." Jeffers, basing himself on his own observations and on the notes made by Bushell during the action, said that the attack was "unprovoked and unexpected," and that at no time had any Paraguayan flag been displayed. He, in his turn, commended Potts, Lamdin, and Taylor for their actions during the encounter.

The following day the *Water Witch* left Corrientes and started down the Paraná to find the *Pilcomayo*, since it would be up to Captain Page to decide the next move.

The Paraguayan government, meanwhile, was not idle. On February 4 Falcón, the acting Minister of Foreign Affairs, wrote to Marcy telling him of the incident and stating that "the . . . attack of the Commander of the Water Witch, whose name we do not know, was gratuitous and without provocation." He went on to explain that the steamer had entered a channel which was definitely Paraguayan water. He sent along a map (quite different from the actual geography, fabricated to fit his explanation) to illustrate the river and the route of the *Water Witch*. He went on to say that no country allows the sketching of positions, etc., particularly those of a military nature, and that Paraguay was in special danger at the present time. He presumed that the United States government would punish the commander of the steamer for compromising U.S.-Paraguay relations and insulting the Republic.[16] Many years would pass before this affair was completely forgotten.

About the same time Falcón was writing his letter the men on the *Pilcomayo* were susprised to see the *Water Witch* following in their wake. Both ships steamed on to Paraná where they anchored and Page and the others came aboard to hear the whole story.[17] Page was obviously enraged and wrote immediately to Dobbin about the "most unprovoked, unwarrantable, and dastardly attack." He had, he said, permission from the governor of Corrientes to explore that part of the Paraná, adding, "I had never heard that Paraguay presumed to exercise the power of preventing its Navigation." (Though he would not admit it, Page did know, as pointed out previously, that the river was a "sore spot" between Paraguay and Argentina.) He then went on to insist that the Paraguayans knew the purpose of the expedition and that the *Water Witch* had no intentions of attacking Paraguay. Yet they had hulled the vessel ten times and killed an American. Page concluded by saying that he was sailing immediately for Montevideo to see the Commodore and make arrangements for some sort of satisfaction at once.*

But even in this moment of crisis Page did not lose sight of his purpose for being where he was. He told Dobbin that if the flagship was not in La Plata waters when he got there he would use the time while waiting for it in surveying around Martín García Island. He ended his long letter with the interesting news that a Brazilian fleet consisting of eight steamers of war, four sloops (totaling together about ninety guns), and ten transports carrying a thousand soldiers was anchored at the moment in Paraná. Much suspicion had been aroused[18]

> . . . by the appearance of this force in the rivers, in connection with a strong land force, said to be at points convenient; and also in connection with the present Brazilian force, now in the "Banda Oriental." These Provinces have an idea that Brazil is looking ultimately to the annexation if not conquest of all those countries on the East of the La Plata, Paraná and Paraguay rivers. The force she has now in these waters and in the neighboring State of Uruguay certainly look to something beyond the simple settlement of her differences with Paraguay.†

In the orders he had received from Secretary Kennedy at the outset of the expedition Captain Page had been told that should any violation of American persons or property occur, he should "seek reparation or restitution by persuasive yet firm measures; and you will not resort to force unless in the last extremity, and when no doubt can exist that right and justice are on your side."[19] So far as Page was concerned this was definitely a "last extremity," and there was definitely no doubt about right and justice being on the side of the *Water Witch.* It was with a view to doing something about the situation that he boarded the vessel, leaving a few men with the *Pilcomayo* in Paraná, and sailed for Buenos Aires, which he reached on February 8.[20]

As soon as he arrived Page wrote a letter to James A. Peden, who had replaced Pendleton as U.S. Minister in Buenos Aires, giving him the details of the Itapirú incident. Peden immediately made an investigation, questioning Jeffers and Page's son, "an intelligent youth, fourteen or fifteen years old." (This is the only indication the boy was on board at the time.) Peden found their accounts "clear and in agreement." The canoe had carried no flag, nor was the bearer of the printed paper dressed as an officer. Furthermore, the ship could not have stopped between the time the blank cartridges were fired and the first fatal shot came, even had she intended to. Nor did the *Water Witch* fire until she was fired upon. These and other facts Peden set down in a

*To the Secretary of the Navy Page insisted, "[With some extra guns and men] I shall feel confident of the ability of the Water Witch to avenge the outrage which has been perpetrated on the Flag of the United States. . . ." Washburn found this remark extremely ironic: "Commander Page was indignant," he wrote. "The flag of his country—alas that it should be said! the flag he himself afterwards betrayed, and endeavored to disgrace and humble—had been insulted, and he was eager to avenge it so promptly that it should ever after be respected in those waters."

Curiously, the two protagonists themselves were not present at the offending incidents (Clement's and the *Water Witch*'s): they both had to protest insults directed at others. In this situation people sometimes react more strenuously than they would for themselves.

(Ponko, 124; Washburn, i 374.)

†Because of the Apá-Branco boundary dispute (n., p. 30), Brazil sent this force—actually some 20 ships and 5000 men—up the Paraná River to intimidate López. He quickly reinforced the fort at Humaitá. The fleet did not reach Paraguay until February 1855—and López simply warned it to keep its distance. This substantial threat—simultaneous with the *Water Witch*'s approach to Itapirú—must nevertheless have increased the tension throughout Paraguay's military establishment, but it was scarcely mentioned as a mitigating factor. (Warren, 187; Page, 45, 109, 313.)

letter to Marcy on February 10. He also enclosed copies of accounts which appeared in the newspapers, including the *Tribune* of that day, which claimed the Paraguayan attack was justified because the *Water Witch* was in Paraguayan waters. The general opinion, however, was that some satisfaction should be required of the Paraguayan government, which was "in nature and character essentially the same as that of the Dictator Francia."[21]

Ignorant all this time of what had been transpiring, the State Department had named Louis Bamberger to replace Hopkins as consul at Asunción. He was in Buenos Aires at the time, and on March 21 the Paraguayan Consul General in that city sent him a letter saying the action at Itapirú had been a necessity on Paraguay's part and that the country still wanted to favor the United States, but that it was difficult to deal with men like Hopkins and Page.[22]

Page, meanwhile, was having his own difficulties in dealing with Commodore Salter.* Within the next two months he visited the squadron commander four times, each time urging some action against Paraguay. First he asked to borrow additional guns for the *Water Witch*, but Salter refused to lend them. Then he offered to tow the *Germantown* up the river to bombard the Itapirú fortifications. This, too, Salter declined to do.† Page finally wrote to Dobbin that Salter did not have United States interests at heart and had refused to do anything. "The Fort which fired into the Water Witch could have been knocked down, by this time; the insult would have been avenged; the Government at home relieved of the embarrassment which it may now experience, and we should then be on an equal footing for negotiations."[23]

Peden also spoke to Salter, and they agreed that it was best to wait for instructions from Washington. In two letters to Marcy Peden summed up the whole situation. The United States, he said, had three causes of complaint against Paraguay: the expulsion of the mercantile association of Hopkins, the refusal to ratify the treaty, and the attack on the *Water Witch* which also involved the right of navigating the Paraná and the payment of some indemnity to the family of Samuel Chaney.[24] "Some think the 'Germantown' sloop of war should be sent to bombard the fort which fired on the 'Water Witch.' I think not." The other two grievances would not be solved by bombarding the fort. He felt that the government would find a way to right things.[25] When Brazil got her way Page said, "It shows the value of having a force to back up your demands."[26]

*William D. Salter (1794-1869) had joined the Navy in 1809—and obviously survived a long career. After this episode he commanded the Brooklyn Navy Yard, 1856-59.

Charles A. Washburn (1822-89), formerly a newspaper editor, as first U.S. minister to Paraguay, 1861-68, was dealing frantically with Solano López' toppling regime; he too wanted naval strength nearby and had similarly been snubbed. Washburn judged Salter's failure to respond inexcusable. "He seemed to have adopted the ideas which have long been prevalent in the American navy, that in time of peace our ships of war are sent to foreign stations for the convenience and pleasure of the chief officers . . .; that the flag officer [Salter] of the squadron is to take no responsibility upon himself, but to make his flag-ship a pleasure yacht. . . ."

(Washburn, i 375; Warren, 245.)

†As a sailing sloop the *Germantown* would not have been able to negotiate a small river easily. The fleet in fact had no ship with draft shallow enough to enter the stream, said Ammen, but Page knew the river was—if only momentarily—high. (Ammen, 277; Page, 313.)

Gaucho.

9. THE URUGUAY AND ARGENTINA RIVERS

For two months after the Itapirú incident the *Water Witch* rode at anchor in Montevideo while Captain Page made his unsuccessful attempts to arouse Admiral Salter to action. Even though this was uppermost in his mind Page did not lose sight of the expedition's purpose. He wrote Dobbin, "Anxious and desirous as I am to return home, I nevertheless feel a greater anxiety for the completion of the work."[1] Several rivers, particularly the Salado, remained to be surveyed. An American company in Buenos Aires had a small steamboat for sale at $10,000, and Page suggested that he be allowed to buy it, in view of the fact that the *Pilcomayo* was not working properly—hinting at the same time that an agent from England also had his eye on it.[2]

Some time had passed since he drew any funds (though he had made two withdrawals not previously mentioned of 2000 pounds and $8000), so he drew 2200 pounds to continue the expeditionary work.[3] Page also spent some time cataloging specimens, and in early April sent a box of twenty-three kinds of wood (the box itself was Paraguayan cedar),* a box of iron and copper ores, soils and sands, and "two small boxes of butterflies and insects collected in Paraguay."[4]

During this same period there was the usual continual turnover of crew men, some being discharged, new ones being shipped, and five men deserting, including one who had been on board less than a month.[5] Page had his personal troubles too. An article appearing in the New York *Herald* for February 1, in which the captain of the U.S. brig *Bainbridge*, Lieutenant Commander Charles G. Hunter,† charged that Page had visited López in preference to Hopkins,

*Paraguay boasts about 500 species of hardwood. The quebracho (from Spanish *quiebra hacha*, "ax-breaker") is also an industrial source of tannin for curing leather; it is an important product for Paraguay. (Weil et al., 17; Kolinski.)

†Charles G. Hunter (1813-73) had become a midshipman 1835, lieutenant 1841, and in the Mexican War commanded a

and that he had forbidden his officers to visit the consul or to carry home any dispatches from him, and that he had even trampled the American flag underfoot. Page wrote to Dobbin to tell him that all of this was calumny, pure and simple.[6] Since the *Bainbridge* was in Rio at the beginning of the year it is not hard to surmise where Hunter got his information!*

Toward the end of April Page decided to make a trip up the Uruguay River,† and on the twenty-fourth he left for Martín García Island and the mouth of the Uruguay. The *Water Witch* anchored in the port of Higuerites the next afternoon.[7] Although this port does not appear on modern maps Page remarked that it had fine facilities and wondered why the early settlers had preferred the not-too-distant Colonia to this site. It had, in addition, fine granite quarries.[8] On the twenty-seventh soundings were made off Nueva Palmira. The river here was about five miles wide, bordered on the Entre Ríos shore by low bushes and trees, and on the Uruguay side by sand beaches backed by rolling hills.[9] On the following day they passed Fray Bentos, where the river narrows to a quarter of a mile, and anchored that evening at Concepción (Entre Ríos).‡ This town of 5000 was "prettily situated on high land" about a half mile from the river.[10] As a sign of welcome and good will the officials sent beef and vegetables for the crew. Gestures such as these led Page to write later, "Politeness and hospitality marked the conduct of the people of the country."[11]

On May 2 the *Water Witch* moved nineteen miles north to Paysandú, population 2388, and the only town on the Uruguay side of the river for a hundred and thirty miles. Surrounded by grasslands and scattered trees, it had become, because of the civil wars, "a forlorn, dilapidated place."[12] A number of townspeople visited the vessel before she left the next day for Concordia.[13] As the steamer ascended the river the banks on either side presented a striking contrast in scenery. The left bank was high, rolling country, of such beauty that, in Page's opinion, it far surpassed that of the Paraná or the Paraguay rivers. The right bank changed from flat wooded country to "undulating grassy hill-sides, gently descending into meadows of surpassing verdure."[14] Twenty-five miles above Paysandú was an estancia five square leagues in area, supporting 18,000 head of cattle. It belonged to an Englishman named Campbell. The surrounding region was filled with "quiet pastoral landscapes," on which there was much mimosa.[15] Higher up the river the Entre Ríos side became a "a beautiful high grassy plain fringed with bushes," with an occasional "luxuriant growth of palms."[16] There was little cultivation, but it was fine country for cattle, horses, and mules. This was the "home" of the well known jerked beef.§[17]

The *Water Witch* arrived at Concordia at 4:58 p.m. on May 8. This town of 1500 was a shipping center for the Misiones region farther north.[18] The "Salto Grande" which Page was

double-screw iron steamer. After one engagement he accepted the surrender of a coastal fortress—and was promptly court-martialed for not awaiting the arrival of his senior officer. Ammen reported that his captain, furious over the *Water Witch* incident, wanted to go upstream to "set things right." Upon Salter's veto Hunter—to show his ire—left the squadron and sailed home without orders, for which he was promptly dismissed from the Navy. He wrote the newspaper letter a few days later. In the Civil War Hunter returned to duty and for firing on a British blockade runner was again reprimanded. (Ponko, 126, 245; Ammen, 277; *National Cyclopedia of American Biography*, ix 186.)

*Hopkins, of course, was extremely active in Rio de Janeiro; see end of Ch. 7. Still only 32, he plunged into steam transportation schemes throughout the La Plata area, including railroads, wharves and docks, and other projects. He was also editing an English-language newspaper and wrote and translated extensively in English and Spanish on matters of national importance. In 1869, upon the fall of Solano López' government, he had what must have been the sublime pleasure, as Argentina's representative to Asunción, of proposing a provisional government for Paraguay. See also n., p. 85. (Ynsfran, 250-55; Peterson, 259-60.)

†Guaraní *uru*, bird, *guay*, land.

‡Concepción (Concepción del Uruguay) is in Entre Rios province, Argentina, between the Paraná and Uruguay rivers.

§From Quechua *ch'arki, charqui*, dried beef; meat cut into strips and dried in the sun.

soon to encounter made river navigation to that area impossible. After only a brief stay the steamer moved three miles north and to the other side of the river to the town of Salto, where it anchored in twenty-four feet of water on the morning of May 10.[19] Since this was the farthest point the vessel could conveniently ascend, Page set out with a group in a small boat the next day to see the "falls."* It took the rowers five hours to pull up to the site of the rapids.

The river at this point was about a third of a mile wide. Across it was a ledge of rocks, reddish in color, over which the water dashed at "inconceivable force," making the stream a "foaming surface." The banks were covered with climbing plants whose flowers filled the air with a delicate odor. Descending from the rapids the river was filled with rocky shoals.[20] Still, Page mused, if locks could be built somehow, vessels of five-foot draft would be able to ascend another one or two hundred miles above the "salto."[21]

On May 12 the *Water Witch* began her descent of the Uruguay, stopping on the sixteenth near the mouth of the Gualeguaychú River. This territory was somewhat familiar from the former visits the steamer had made in her first months in the La Plata region. After surveying this area the ship moved across the river to Fray Bentos. This proved to be the nicest city on the river. There were close to 14,000 people, and Page noted that "the public and private buildings are handsome and substantial structures." Even though it was in another country, Urquiza had a home here.[22] The expedition spent five days in and around Fray Bentos and then proceeded to the mouth of the Negro River, which is Uruguay's most important inland waterway.[23] Murdaugh and Page went upriver to Mercedes, a town of 7000. It proved a benefit for the expedition because the area around it was abundant in ancient animal remains.† After three days the two returned to the steamer and she sailed for Buenos Aires, arriving there on May 28.[24]

Page was enthusiastic about the results of the expedition. The *Water Witch* had gone two hundred and ninety miles up the river and found it navigable all the way. Furthermore, it would remain so in any season of the year, though at times it would demand "extraordinary steam pressure." The countryside, though beautiful, was "a spectacle of sad waste" because of the wars, but the land had rich soil and fine pastures and tremendous agricultural possibilities. Rural labor alone seemed to be lacking. There were only "the gauchos,‡ the only laborers of the Riverine Provinces," and they were not interested in farming.[25]

Page had planned to survey some of the small streams running directly into the La Plata River when an accident changed the course of things. The crosshead§ of the engine broke, forcing the steamer to remain at anchor in Buenos Aires. Page described the mishap to Dobbin, saying that while it could be fixed, the crosshead should be replaced by a new one, which he hoped could be received in Buenos Aires "in less than two months." He then told the Secretary that he had contracted for the use of a small steamer called *Yerba*, and with it he intended to explore the Salado River. He would give, he said, the *Pilcomayo* in payment. When the crosshead was fixed,

*The "Salto Grande" ("Great Falls") was misnamed, said Page; it was only a rapids. (Page, 326.)

†The La Plata area is uniquely recognized scientifically as a land of primeval life. Pineapple (*caraguatá*, a wild species, grows here) and maize (corn) may have started in Paraguay: *abatiguaniba* (wild maize) is still used here for bread. Many other plants are ancient or found in few other places.

Animal life too is primitive. The lungfish is here; in water it must surface to breathe, and when a lake dries out it burrows into the mud to "hibernate." The opossum, armadillo, and ground sloth (and extinct toxodont) are believed to have originated in South America, migrating northward. Only the toxodont failed to reach North America.

Darwin in 1833 temporarily left the *Beagle* (see n., p. 80) and traveled several hundred miles to see "giants' " bones said to be embedded in the Paraná's banks; he found the fossils but they were too fragile to remove. Murdaugh here discovered glyptodon and megatherium bones—not in the ground but at the home of an Englishman living in Mercedes.

(Page, 145, 329; von Hagen, 193-94; Weil et al., 17, 196.)

‡See n., p. 36.

§A block connecting the piston rod of a steam engine with the connecting rod.

and while he was gone, the *Water Witch* would survey the La Plata waters, particularly around Martín García Island.[26] At the same time he wrote to his friend Maury, showing some of his old enthusiasm and telling him that when the work of the expedition was completed, there would be a "creditable portion" of work to lay before the public, and it would most certainly "enlarge the bounds of geographical knowledge."[27] The *Yerba* anchored next to the *Water Witch* on June 30, and men and equipment were transferred to her. On July 2 Page, Powell, Murdaugh, Stump, Lamdin, and twenty-one men left at 3:00 p.m. for the Salado River.[28]

The steamer *Yerba* was one hundred and twelve feet long and had a twenty-six inch draft.[29] At this size she could conveniently explore some of the smaller streams that emptied directly into the Paraná. The first days she entered a few and went far enough up the Capitán to visit the village of Baradero. The stream or *riacho** was three hundred yards wide and bordered mostly by marsh grass.[30] By July 7 the men had reached San Nicolás, where they investigated riacho Number Five. In succeeding days a few more small streams were explored, and on the eleventh the *Yerba* arrived at Santa Fé.†[31] In the three years that the *Water Witch* had been in the La Plata region (1853 and 1855 partial years), the population of the province of Santa Fé had tripled from 4000 to 12,000. Half of that number was in the city of Santa Fé itself. The former fear of the Chaco Indians was decreasing, while trade with the interior provinces was growing. The region around Santa Fé furnished 50,000 mules a year for the Alto Peruvian market.‡[32] This interest in the interior was heightened tremendously by the projected expedition of the *Yerba*. People were excited by the prospects of what navigation to the interior could mean, and the atmosphere and enthusiasm in Santa Fé was very much like that of an American frontier town in the days of the opening of the west. Some said the steamer could ascend perhaps forty-five miles.[33] Page was hoping for a great deal more than that.

On July 13, with the governor of the province and his family and friends aboard, the *Yerba* left Santa Fé and entered the mouth of the Salado River. It was fifty yards wide and the water from twelve to eighteen feet deep. To Page a narrow, deep river meant a very long one. His hopes were high. Eighteen miles upstream, at Santo Domingo, the governor and his group disembarked and the steamer continued her winding way.[34] The course of the river was so meandering that the vessel traveled seventy-five miles to reach a point twenty miles distant from the starting point! Progress was slow and poor visibility in rainy weather caused the steamer to strike the banks now and then. The river varied greatly in width, but in the bright, clear days that followed the vessel was able to move along easily. On the sixteenth they reached Paso Yrionda, where the men saw a number of Santa Fé soldiers. This was the last outpost.[35] When they passed this point Page wrote, "All civilization is behind us; we are entering the undisturbed possessions of the Chaco tribes."§[36]

It really seemed a sort of frontier, for suddenly Nature came into her own and plant and

*From the Spanish *rio*, river; *riacho*, little river (also *riachuelo*, brook). Ammen defined riacho as a lateral or inferior channel. (Ammen, 257.)

†According to Hutchinson Santa Fé thought itself the quintessence of refined tropical society. The siesta was observed 11 a.m. to 3 p.m., with the "maté siesta" continuing to 4. Only "madmen" went for a walk 6-7 a.m. (the best time) and during maté siesta no one was on the street "except dogs and Englishmen." (Hutchinson, 243-44.)

‡Peru's broad, high plateau continues into Bolivia and northern Argentina as the Altiplano or "Alto Peru," an elevated plain some 500 miles long and 80 wide. Before the conquest the Guaraní long had commerce with the Altiplano Incas.

§The Chaco was long recognized as home for a variety of untamed natives, among whom were the Chiriguano, Guaieni, Guaraní, Guaycurú, Itatin, Makka, Mbaya, Orejon, Querandi, Toba, Tupi, and Xarayes, some having resisted the Spaniards fiercely. Page's expedition made no attempt to investigate them. They have not been studied systematically until recent times; a few of these tribes are now extinct. (Kolinski.)

animal life was wild, abundant, and beautiful. Rolling land with scattered trees, cliffs some fourteen feet high, expanses of grass and bushes, and "fine land well wooded down to the bank" followed one another as the steamer moved onward.[37] Animal life was everywhere present:[38]

> It is the domain of the jaguar, the shielded armadillo, ducks, geese, flocks of the black-necked swan, plover of different species, partridges, pigeons of extraordinary size, the guanaco,* the ostrich, the hare, the deer, the clumsy capibara, while the waters teem with fish and the air is darkened by flocks of the small white gull.

And all of this only thirty-three miles in a straight line from Santa Fé! By river the *Yerba* had come a hundred and fifty miles. The stream itself was still twelve to twenty-nine feet deep.[39]

In the next two days the landscape changed enormously. It became a lagoon-dotted pampa,† "marshy land with hundreds of waterfowl," only to change to grassy flats filled with trees again.[40] By July 23 the *Yerba* was well into the interior, seventy-five miles by land and three hundred by river. A small stream called La Cruz was investigated and the men came upon a great salt bed, one of the many in the region which gave the river its name—Salado.[41]

Two days later the steamer came to a point where another large stream entered the Salado. The log recorded: "The river divides into two branches of mostly equal size, that from the eastward is the broadest a little but has little or no current while that from the north and west has considerable." The water in this stream was "yellow and salty."[42] The *Yerba* took the left branch since the current indicated it as the main river, but the next day Page decided to go no farther. The water was decreasing, there were many fallen trees in it, and the river itself was narrowing—so much so that the steamer had to back down to the La Cruz to turn around. So rapidly was the river falling that on the way down the men saw an island where there had been none before, and the steamer was grounded several times.[43] Until they reached Santa Fé on August 6 there was nothing of incident except the shooting of a large jaguar crossing the river. It measured five feet two inches from nose to caudal bone.[44]

For several days after their return to Santa Fé the men were working on the *Yerba* repairing the wheels, painting, reloading fuel, and other chores. Murdaugh was sent on a special mission to Paraná where he remained for one day.[45] During this time the specimens of the various birds caught and killed on the Salado were prepared. Among them were a spotted duck, a bronze-colored duck, a blue-wing duck, a black drake and duck, male and female spoonbills, several divers, a flamingo, a swan and a crane.‡[46] Page meanwhile was preparing to make an overland trip into the interior with the idea of continuing the exploration of the Salado in a smaller boat. He was convinced it was navigable far beyond the point the *Yerba* had reached. On August 12 Page, Murdaugh, and a sailor named Thomas Cornelius set out with Colonel Rodríguez, the military commander of the province, on the first leg of a journey to the distant town of Córdoba.[47] The *Yerba* was to make a few more exploratory trips and then return to Buenos Aires before the expiration date of the contract.

On the same day that Page left the *Yerba* got under way and sailed down the river to the Paranacito, a stream which it entered. Two smaller tributaries were explored, as was the rivulet itself for four days. Next the Gualeguay River was ascended and found to flow through fine

*The guanaco (or huanaco), about 3 feet high, resembles a deer but is related to the camel; it is the wild ancestor of the llama.

†A Quechua word meaning "flatland"; an extensive grass-covered plain or savanna.

‡In all, the expedition collected some 78 species of bird, and although Page furnished the Guaraní name of each, he referred to them in his writings in general North-American terms. Ponko nevertheless credited him with "the best natural history collection of that region." (Harrison, 198; Ponko, 132.)

pasture lands filled with horses and cattle. In succeeding days the Pabón, Paraná Guazú, Capitán, and Luján rivers were all entered, explored, and mapped. On August 29 the *Yerba* anchored alongside the *Water Witch* off Martín García Island.[48]

Fort Bourbon, or Olimpo.

10. MOMENTOUS DISCOVERIES

After he had returned from his Salado trip in the *Yerba* Captain Page wrote to Dobbin to tell him of the exploit. They had gone, he said, three hundred and sixty miles from Santa Fé. The Salado River had never before been navigated to that point, mainly because of the fear of Indians. "The region of country," he wrote, "through which we passed is beautiful in scenery. . . . It only wants the hand of civilized man to transform its wilderness into a garden."

The governor of the province, Señor Cullen, looked upon the trip "as the opening of a new era in the prosperity of those provinces through which the Salado passes." Page felt it was very important to establish the fact of navigability as far as possible, and he intended to continue the trip from another point inland and higher up the river. He had asked people about the waterway, but "there was no information to be obtained because the people of Santa Fé are as ignorant of the subject as though the Salado were a river of India."[1] The only solution was to go and investigate it himself.

On August 12 he set out with his companions. They traveled first to a military post twenty-two leagues from Santa Fé. At this site, called Romero, he was given an escort to accompany him into the province of Córdoba, because of the possibility of encountering Indians.[2] The group traveled across an unbroken plain for forty-seven miles to Tío, a village of thirty white-washed adobe houses. The ride from here to Córdoba was long and tiresome. Small military posts were spaced from twelve to twenty-four miles apart where a change of horses could be made, but otherwise they offered only "an empty house, scanty fare, hide cots without bedding, and indifferent horses."[3]

To Page's surprise the country between Tío and Córdoba was sprinkled with wheat fields and neat houses. Córdoba itself impressed him favorably. He called it "The Athens of La Plata."

Situated on the banks of the Primero, it was the center for the northwest provinces. There were ten churches and a cathedral, and several convents with walls twenty feet high. The abundance of their artwork undoubtedly made him compare this city of 15,000 to Athens. When they heard of his trip the people of Córdoba were not too enthusiastic about it. They felt it might take away trade. To allay this fear Page suggested they purchase Conestoga wagons* and modern agricultural implements from the United States to improve their own transportation and work methods.[4]

From Córdoba Page and his companions turned northward on August 25 for the rest of the long journey to Santiago del Estero. Along the way were twenty-one posts for changing horses, but they usually had to wait an hour at each one. The country through which they passed was low sierra; sheep dogs were common. Except for one village called San Pedro, where they saw apple and peach trees, most of the region was desolate—sandy and gravelly soil in which only scrubby bushes grew. Once they came upon a gigantic cactus with "rich, lucious yellow fruit."† In general, there was little of attraction. So scarce were people that one night when they met a train of ten wagons the two groups got together for an evening of revelry—such as a barren plain and limited supplies afforded. Page discovered on this trip that if he drank yerba maté he could ride almost all day without food or hunger.‡ [5]

In early September the group arrived in Santiago del Estero. This town of 5000 had four main buildings—three adobe churches and a government house of brick. The region around the town was full of hardy fruit trees, and there was an abundance of cotton. The climate was especially invigorating, and one day when Page thought he was ill the people told him that was impossible; sickness was unknown there—"no physician or apothecary" even living in town. When Page explained to the province governor, Manuel Taboado, his interest in the Salado, the Governor enthusiastically supported his plans, and offered his own boat for the trip. He used it occasionally on the Dulce River which runs by Santiago. Since the Salado was some fifty miles east, the boat had to be taken overland by ox cart. On September 8 Page and his party, the Governor, and the Governor's brother Antonio, who knew Quechua,§ started for the Salado, accompanied by eighty soldiers as protection against the Indians.[6]

The trek to the river was through "fine country, forested and flowered." Page noted eight different types of bee. When the river was reached the eighteen-foot boat was launched under the Confederation flag and started down the river. People along the way helped by chopping away trees and bringing gifts of food. The sight of a boat on the river was so unusual some of the country folk along the way thought it was a large animal or Indians in some novel disguise. The travelers descended to Matara, a little village of one church which had had no priest for two years. The inhabitants held a feast to celebrate the arrival of the boat and Page noted, as Ammen had in Paraguay, the absence of drunkenness. The people were "primitive but not vulgar."[7]

*From its place of origin, Conestoga, Pennsylvania; it was a fairly sophisticated covered freight wagon with upturned ends (so cargo recenters itself over rough terrain). The natives were using the *carreta*, a primitive two-wheeled mule- or ox-drawn cart; designed for moving tree trunks, its enormous wheels up to six feet in diameter gave it high clearance for backwoods trails. The people of Córdoba actually ordered $50,000 worth of merchandise; it was filled in Boston and arrived in Rosario about two years later. (Page, 350; Kolinski.)

†The natives made a delicious jelly from the fruit. (Page, 355.)

‡Outsiders are not usually so impressed with the tea. The traditional serving method is to steep it in a gourd (the "maté") and insert a pipe or "straw" (*bombilla*) often made of silver with a screen at the bottom to trap the particles. This was passed around in a group. Wilcocke found it an aperient and diuretic, habit forming, and conducive to disorders similar to alcoholism. "Like opium," he said, "it produces some singular and contrary effects; it gives sleep to the restless, and spirit to the torpid." (Wilcocke, 495-96.)

§Quechua is the language of the Quechuas (Quechuan, "plunderers"), a native people of central Peru; it is also spoken in Bolivia, Ecuador, Chile, and Argentina.

While exploring in this region Page one day joined a group of soldiers chasing a band of Indians. The Indians outrode them, but the pursuit, which lasted thirty-six miles, ended in the recapture of two hundred horses and two hundred and fifty horned cattle. About two weeks later, at Paso Mistal, the expedition came upon another band of raiding Indians, and again there was a chase.* At one point a soldier was killed in a skirmish. An Indian encampment, recently deserted, was discovered. Page noted that one tent had its poles bound with the hair of a white woman. The cloak and headdress of the cacique were found and taken as prizes. Page and Murdaugh had to add sleeping in the open boat all night to their list of inconveniences during this part of the exploration.†[8]

In early October the return trip to Santiago was begun and the party arrived there on the eighteenth. Page sent a letter to Dobbin describing the expedition and its enormous consequences. He had established the navigability of the Salado for eight hundred miles. "Never, until now, has it been known to be navigable." True, there were some obstructions from overhanging and fallen trees, and from some water growth, but these could easily be removed. The tremendous market made available would make the work well worth it. The river could replace the clumsy ox carts which had to travel from three to nine hundred miles. The discovery was judged really important: "The enthusiasm of the Governor and principal men is truly surprising," Page said. "It could not be greater had this exploration developed a gold mine in their midst." Page was treated as a hero and "delicate attentions" were showered upon him everywhere. He concluded by writing that he intended to see about the upper portions of the river at Salta and would have to abandon his plan to descend the Pilcomayo River "because of the revolution now going on in Bolivia. . . ."‡[9]

A grand ball was held in Page's honor on October 24, and three days later he left for Tucumán.§ Once again the officials received him most cordially. He was taken on a tour of a sugar plantation one day, and on another he went into the country where he was impressed by the snow-capped Catamarca Mountains and the great forests. On October 30 a dinner was given at which the governor, José María de Campo, and the minister of state, José Posse, and other distinguished guests were present. In a speech for the occasion three great epochs in Argentina's history were claimed: Independence, the downfall of Rosas, and the discovery of the Salado's navigability. Whether serious or not, it was a great compliment and must have given Page and his men great satisfaction and joy. One of the minor pleasures of his visit was the discovery of an English doctor in the town. It was good to be able to converse in English again.[10]

*After his return to the United States Page apparently described these chases to the Secretary of the Navy as if they involved some physical danger. No actual Indian encounter occurred, "not considering it my duty to be speared," he wrote. There were no details about the woman's hair (below). (Page, 378; Ponko, 126.)

†As Charles Goodyear had just patented his vulcanization process (1844), Page was also reporting on rubber impregnated cloth. The "shawl and poncho [were] no effectual protection against damp grass and the chilliness of the night," he wrote on one occasion. "I shivered; for my India-rubber poncho was a good protection from the rain, but a very poor one from cold," on another. But after the Indian chase, when he spread his "India-rubber blanket on the pampa" he awoke "refreshed, as if my night had been passed in the most luxurious apartment. . . ." His final analysis:

"The privation and exposure we experienced throughout this exploration, not only of the Salado, but of the country, somewhat in the interior of the provinces, were calculated to try our physical endurance, and test our zeal and energies even in a cause of such deep interest and importance. We nevertheless found ourselves relieved from the fatigues of the day, after a night spent under the soft grass of the pampa, with the canopy of heaven our vaulted roof. The rain would descend with tropical force, but our India-rubber ponchos seldom failed to secure us refreshing sleep. The dawn would find us, though wet, prepared for the toils of the day. . . ."

(Page, 329, 381, 386; Ponko, 126.)

‡Between 1848 and 1855 alone Belzú's regime (n., p. 74) faced some thirty or forty political revolts.

§Tucumán was formally founded by Spaniards in 1542. (Wilcocke, 264-65.)

On the morning of November 4 Page and Murdaugh left for Salta. As they moved north the arid mountain desert gave way to verdant, fertile, well watered valleys. They were full of fruit trees and wheat fields and villages of soundly built homes. Then came the exciting experience of crossing the mountains and proceeding through precipitous mountain passes. It took the two men six days to reach their destination. It was a difficult trip and, as Page tersely remarked, it had "no tourist accommodations." The journey had its reward, however. Page found Salta a remarkably fine town: "It is ahead of all other towns of the Confederation except Rosario." This community of 10,000 had two-story stucco houses and a great deal of commercial activity and enterprise. This led Page to see once again the value of the Salado River: it would shorten the land carriage by four-fifths.[11]

To his surprise Page met Hilarión Ortiz, one of the four Bolivians he had taken down the Paraguay River from Coimbra, Brazil.* Through him he also met General Santa Cruz,† the famous Bolivian ex-president. Santa Cruz had been offered a position as Minister to France by President Belzú,‡ even though he had been refused admission into Bolivia itself. The General declined, and he and Ortiz were awaiting a political change in Bolivia which they thought would occur soon, since it was rumored that Belzú's son-in-law, Linares,§ was already attempting to overthrow the government.[12]

Page and Murdaugh remained in Salta a week, leaving on November 14. Two days later, at Las Piedras, they parted company, Page heading directly south and Murdaugh turning slightly east to follow the Salado down to the point where the boat had been launched, in order to see if its upper reaches in this area were navigable. They agreed to meet in Santiago. On his way there Page met the first stage coach to make the trip from Tucumán to Salta. He managed the Tucumán-Santiago trip in two days, riding a hundred and five miles in one day. Three days later Murdaugh arrived with the information that the part of the river he had investigated was full of sand banks which would make any navigation very difficult.

The men continued to Córdoba. While trotting through town Page was stopped by a policeman and fined two dollars for riding too fast in the city! The incident amused Page so much he paid quite willingly. Five more days of travel brought the men to Rosario, where they were guests of the British consul, Mr. Dale. On December 12, after an absence of a little more than five months, Page and Murdaugh rejoined the *Water Witch* in Buenos Aires.[13]

Any odium the expedition may have suffered in Argentina because of the Itapirú incident had vanished completely, dispelled by the joy and enthusiasm caused by the Salado expedition, now already a nationally publicized event. The newspaper *Argentino Independiente* of November 10 said:[14]

> According to the information concerning the Río Salado which has been given to us, in various extracts of letters forwarded for publication by the Minister of Government, Don José Posse, we can say that the navigability of the Salado is now an established fact. We are lost in wonder and admiration when we consider this achievement, and the social and commercial revolution it is destined to work out in the condition of these provinces.

*See n., p. 32.

†Andrés Santa Cruz (1792? -1865) had been active in Peru's revolution against Spain, 1820-23. He became president of Bolivia 1829-39, still hoping to reunite Peru and Bolivia. Despite a long, stable administration he was overthrown and exiled to Europe. He was succeeded by Ballivián (n., p. 14).

‡Manuel Isidoro Belzú (1808-66), a Bolivian general, was president 1848-55. He was assassinated and succeeded by his son-in-law, Linares.

§José María Linares (1810-61) would become president of Bolivia 1857-61. The country's first civilian president, rich and distinguished, Linares wanted so honest a government he ruled ruthlessly to get it. He was forced into exile and died the same year.

The *Comercio* of Salta added that the people were "grateful to the North American government," and that Lieutenants Page and Murdaugh had left with them "remembrances that will immortalize their names."[15] Page sent these extracts to Dobbin along with a full account of the trip, and gave as final proof of the exploration's success the fact that officials were "already talking about moving the town of Santiago to the banks of the Salado."*[16]

Strangely enough, while the people of Argentina were rejoicing over this discovery another one of perhaps even greater magnitude had been made which was to their disadvantage.

About the time that Page and Murdaugh were starting their overland trip the *Water Witch* began a series of soundings in the La Plata, especially around Martín García Island. It was a long, laborious job. In addition to the crew there were always two officers at work. One had a chart on which was projected the course, distance, width, and depth of the river, as well as the topography of the bank; the other recorded all of this, with further observations. Soundings were made every five minutes in deep water, and as often as possible in shallow water or shoals.[17] Except for two two-week periods in Buenos Aires and Montevideo, this work of the *Water Witch* went on for almost three months, but it brought surprising–even startling–results.

Two incidents occurred during this time to break an otherwise regular routine. The first took place August 22 when the commander of the fort on Martín García signaled the *Water Witch* and asked to be brought aboard. He said the garrison on the island had mutinied and he needed help. Lieutenant Jeffers and twelve men landed with one hundred rounds of buck† and ball cartridges. One of the officers had been wounded eight times, and eighteen men had fled to the other side of the island and pushed off in a whaleboat. They were pursued by the light boats from the steamer, but because of their head start and the darkness they escaped. The men from the steamer left the island as soon as order had been restored.[18]

The second incident happened in late October. One day while the *Water Witch* was making her usual soundings a grounded Spanish man-of-war was encountered. It was "Her Catholic Majesty's Schooner *Cartagena.*" The *Water Witch* tried to pull her free, but it did not have enough steam. Two more attempts were made the following day, but the schooner held fast. She was in increasing danger, as the waves and wind were combining to capsize and swamp her. On the third day the steamer finally succeeded in pulling the schooner free, and then towed her to Martín García Island. The next day the commander of the *Cartagena*, Pedro de Aubarede, Teniente de Navío, came aboard the *Water Witch* to offer his profuse thanks for the assistance given. Certainly the schooner "would soon have become a wreck but for the timely aid of the Water Witch."‡[19]

It was just around this time that the summation of the surveyings showed evidence of an important discovery. The log records simply, ". . . prooved [sic] beyond doubt that a new channel exists in the La Plata estuary."[20] This channel was two feet deeper than the one used by all vessels ascending the Paraná and had been "heretofore unknown."[21] Page pointed out the significance of its discovery:[22]

> The Island Martin Garcia, now in the possession of Buenos Ayres, commands the old channel most effectually; through the new channel a vessel may pass at the distance of one and a quarter miles from the Island. The new channel passes between Martin Garcia and the Banda Oriental. . . . [T]he discovery of this

*Santiago, capital of the department of Santiago del Estero, is still on the Dulce River. (Page, 362.)

†Buck is coarse lead shot, intended to wound rather than kill. Ball is the conventional lead bullet.

‡For this gallant service Jeffers was later presented a sword by the Queen of Spain. (*National Cyclopedia of American Biography*, iv 281.)

> new channel deprives Martin Garcia of the importance attached to it, previously, as a military point, commanding the navigation of those rivers . . . and the new channel becomes the common boundry [sic] between two foreign states.

This was another feather in the expedition's hat, and when Page sent word of it to Maury he told him to publicize it at once and to make sure the United States, not England or France, would get the credit for it.[23]

During the month after this discovery the *Water Witch* made some final surveys of rivers visited only briefly before, and particularly checked the places the *Yerba* had not time to investigate thoroughly or at all. The Pabón River was revisited to the village of Ibicuí. The Colastiné and Santa Fé streams were investigated, and Rosario was visited for the last time. It was just two days before Page and Murdaugh got there.[24] They were probably unaware, as they went downstream to Buenos Aires, that a vessel going the other way was carrying a man who was to be a new character in the last act of a drama in which they had just participated—who was to give one of the play's subplots a reasonably happy ending.

Louis Bamberger was the new United States consul to Paraguay. He arrived at Asunción on November 23, and two days later Minister of State Falcón called on him and told him that President López wanted to meet him. When he went for the interview López received him kindly and expressed his good wishes toward the United States government and its citizens and assured the new consul that all Americans were most welcome in Paraguay. He stated again that the trouble with the former consul was of an entirely personal nature, and that he hoped things would work out well. He also added that even the troubles with Brazil seemed to be at an end, and that Francisco Solano López was going to Brazil as Minister Extraordinary.*[25]

Extraordinary indeed! The whole expedition had been that from beginning to end. Certainly the eyes of any adventurer would seem something of a nimbus about the *Water Witch* as she rode at anchor on the gentle swells of the La Plata, awaiting her captain and the last phase of her South American journey.

*Paraguay's troubles with Brazil and her other neighbors were just beginning. To settle the northern boundary dispute (n., p. 30) López reluctantly agreed (1856) to meet with Brazil within six years; it was a delaying tactic for both countries. When he began hampering the latter's shipping on the Paraguay, Brazil with Argentina proclaimed the entire La Plata basin free to all nations. Outwitted, López—on his deathbed—advised his son to settle peacefully with Brazil, but Solano López was not so inclined. Politically isolated, but refusing to surrender control of the river—in fact yearning to dominate the whole La Plata region—Solano López seized a Brazilian steamer (November 1864), precipitating war. When Argentina would not let him cross Corrientes province to attack Brazil (January 1865), he declared war against her as well. Uruguay joined the allies and made the "Triple Alliance" a reality (May 1865). It was a foretaste of World War I. The slow-moving contest of attrition, disease, indecision, and delay ended 1870 with the death of "El Mariscal," killed when he refused to surrender to Brazilian soldiers. Paraguay suffered some 300,000 casualties out of a population of 525,000. (The present-day population of Paraguay is 2.5 million.) (Kolinski, 171; Warren, 187-88, 215-16, 218, 222; Brodsky, 238; Weil et al., vii.)

Maté and bombilla.

11. THE END OF THE EXPEDITION

During the last months of 1855 there was considerable change in the crew of the *Water Witch.* The terms of many of the men had already expired but they had been kept on board. A number of them, consequently, left the ship whenever an opportunity presented itself. In spite of this about forty of the original crew members remained with the vessel until the end of the trip. Page sent Dobbin a list of the crew on December 29, but it was not entirely complete or correct since others joined the ship shortly thereafter, and the list included the names of some persons whom the log records as having left earlier. According to Page's list there were eleven petty officers, six seamen, five ordinary seamen, eight landsmen, five firemen first class, four firemen second class, four coal heavers, and four boys—a total of forty-seven.[1] With the officers and newcomers the total compliment of the steamer was somewhere around sixty.

In early January one last short surveying trip was made off the mouth of the stream called de las Calmas, not far from San Isidro. The *Water Witch* spent her last days in Buenos Aires from January 19 to 23, when she left for Montevideo.*[2] This metropolis had been the main center of supplies throughout the expedition, and with its growing population (now 80,000)[3] it was fast becoming the principal port in the La Plata region. It took ten days to load the ship with the staples she would need. During this time Page wrote to Dobbin that the expedition had done all it could and had completed its task, and he presumed he was to return home.[4]

Even though Page had felt the work completed, from the time he left Montevideo on February 3 he continued sounding the La Plata and the passage out to the Atlantic for several

*Montevideo ("I see a mountain"), capital of Uruguay, owes its name to the "mount," which rises nearby to about 500 feet. In Page's time the top had a lighthouse rising from the remains of a fort. (Page, 33-34.)

days. The first stop, made at Florianópolis on St. Catherine Island, was an unexpected one, necessitated by a leaking boiler. This took three days to repair, and then the steamer headed for Rio de Janeiro, arriving there on February 12.[5] The run from St. Catherine had been made in fifty-four hours at almost eight knots, still good time for the somewhat tired wheels and machinery of the vessel. In addition, Page told Dobbin he intended to return home through the West Indies, to take advantage of the Gulf Stream and to have a southern U.S. port nearby in case the machinery should fail altogether.[6]

The exploits of the *Water Witch* had already become well known, and when she reached Rio she was boarded by well-wishers from English, French, and Brazilian boats. The captains of the *Indefatigable* and the *Thisbe* visited the vessel, and Rear Admiral Hope-Johnstone* of the English ship *Madagascar* came to pay his respects. Most of the time in Rio was spent in resupplying the ship with such items as butter, flour, tea, candles, soap, potash, roman cement,† canvas, and white lead. By February 25 the ship was ready to sail, and under sun and blue skies she started the next lap of her journey.[7]

Another emergency stop was made near St. George, Brazil, to change the position of the buckets on the wheel because one was broken, and to fix the boilers which were leaking again and caked with salt deposits.[8] The next night the *Water Witch* steamed into Bahia. Here she took on water, wood, and coal, and tried to make all the necessary repairs. Little aside from routine affairs occurred during her nine-day stay except that on the day after her arrival the American consul came aboard, and on March 11 the ship was dressed for the birthday of Princess Januária.‡ The heat during these days was oppressive, ranging from the high eighties to the low nineties.[9] The men probably wished more than once that they were back in the temperate La Plata—or better, already home. At 2:10 on the afternoon of the eleventh the steamer got under way again on a smooth sea of very light-colored water.

Six days of sailing brought her to Ceará (now Fortaleza). Here some exceptionally fine coral was found and added to the collection. Two days out from this port, in the area of the equator, schools of small flying fish were seen, along with large flocks of sea birds.[10] It might seem strange that such sights would be of enough interest to be recorded, but on the open sea with nothing but sky and water even small things become significant. Too, after three years of scientific endeavor the observation of nature had become almost automatic.

The *Water Witch* was averaging about a hundred and sixty miles a day in spite of the fact that she was traveling with a hastily repaired crank strap. On March 29 she entered the harbor at Bridgetown, Barbados. Here, in view of paying the crew when they reached home, Page made his last withdrawal from Baring Brothers (he had also received 700 pounds in Rio) of 1100 pounds sterling.[11] The stay in Bridgetown was prolonged to twelve days, due to the efforts of the engineers to try to regulate the engines and boilers which had been giving almost constant trouble. Engineer Potts reported "the immense weight of the wheels has been the cause of fully three quarters of the derangements of the engine."[12] Perhaps even in this the *Water Witch* was contributing something to science and navigation, since it confirmed Page in his opinions about the Morgan wheel, about which he had at first been so enthusiastic and which had proved, in the last months of the expedition, to be more a hindrance than a help.§

*Sir William James Hope-Johnstone (d. 1878) was one of a long family line of distinguished British naval officers.

†A natural cement prepared by heating and grinding limestone nodules into a fine, dry powder.

‡Januária (b. 1822) was an older sister of Pedro II, next in line for the throne. In 1849 she had nevertheless married the Count of Aquila, brother of Doña Thereza (n., p. 42); they eventually lived in England.

§Page had found the eccentric mechanism especially vulnerable in fresh-water streams, where sand and other particles aggravated its abrasion. At sea this was no problem, and Page considered it superior to the common "radial" wheel.

The old established marine shops, like Morgan's, had long opposed the new screw propeller developed by John Ericcson.

On April 10 the steamer left Bridgetown and headed northwest, passing Santa Lucia late that same day. Six days later she reached Port Royal, Jamaica, where a brief stop was made for health inspection, and at 2:25 p.m. the same day she docked at Kingston. At the very same hour, three days later, having taken on the necessary supplies, she weighed anchor and continued her northwesterly course through the Caribbean, rounding Cape San Antonio at the tip of Cuba on the twenty-first, and tossing out her anchor in the harbor of Havana on the evening of the twenty-fourth.[13] Once again the vessel was the object of official attention and was visited by the captains of a French brig and the Spanish flagship. On April 29 the third of three deaths to occur on the trip took place when Charles Turner died. The log records no cause of death but states that at 5:00 p.m. all hands were called on deck, the flag was lowered to half mast, and the body sent ashore for burial. The voyage of the *Water Witch* was an episode in life, and as such it had to endure its moments of sadness. The following afternoon the steamer headed north on her final run.[14]

The rest of the journey was merely a matter of eager waiting, accentuated by the balmy, warm spring days. Cape Hatteras was passed on May 4, and the following day a pilot boarded the vessel at Hampton Roads. The sixth was spent traveling up the Chesapeake to the mouth of the Potomac, where the steamer anchored and Page sent a note to Dobbin asking for instructions as to what to do with all the equipment purchased for the expedition.[15] Page received an answer the same day and the next morning the ship steamed into the Washington Navy Yard from which she had left exactly three years and four months before. The next two days were spent in unloading the ship. Secretary Dobbin came in person to offer his congratulations. When the vessel was cleared it was hauled into one of the yard slips and secured, and the men given their pay. A note of finality rings through the last entry in the log: "Discharged the crew and put the ship out of commission."*[16]

The exploits of the *Water Witch* received widespread attention at the time. The trip was given "highly complimentary notice" before the Royal Geographical Society of London by Lord Ellesmere† and Sir Charles Lyell.[17] President James Buchanan in his first annual message to

(Ironically, Ericcson's 1845 steamer, *Iron Witch*, a paddle-wheeler converted to screw propeller, was a costly failure.) Nevertheless, the screw quickly won in naval applications: it freed the sides of the ship for additional cannon, and, entirely below water level, it was safe from enemy gunfire—as the *Water Witch* demonstrated by taking a hit on its wheel at Itapirú. (Page, 30-31; Ponko, 13.)

*About the time the *Water Witch* returned, said Stanton, "senators were confessing themselves weary of 'very heavy works' on 'bugology' and other 'unimportant and useless branches of natural history' and were inclined to regret the precedent they had set with 'Wilkes' Expedition.' " Congress published only Page's bare report (Wendell, 1856). In an expanded trade edition (Harpers, 1859) Page—not unlike Hopkins—greatly exaggerated Paraguay's commercial potential, apparently, said Harrison, to make his work there seem more important.

"Without systematic training in any field of science and without artistic sense," Harrison noted, "[Page] tried to report like Humboldt: all details were recorded, but no meaningful picture of landscapes as a whole emerged." (The present reader will note that, in quoted passages, Page often comments on the scenic beauties of nature—not really of much scientific value.) Harrison nevertheless rated Page's "local reporting" better than that of any other Latin American expedition.

In December 1856 Congress appropriated money to print Page's charts—and an additional $25,000 to continue explorations in the La Plata region. While preparing the charts—some 56 of them in rough form—Page began to plan the return expedition, which he would command. With only $25,000 to work with, the Navy could not outfit, man, and dispatch a vessel as it had before. Page found an old friend, Robert B. Forbes, a Boston merchant, willing to build a steamer and lease it to the Navy in Argentine waters. The ship was to be the *Argentina*, a three-masted merchant vessel 98 feet long with 16-foot beam and 4-foot draft. Suddenly, however, foreign affairs (see Ch. 12) called Page away from these deliberations.

(Harrison, 197, 200; Ponko, 127-28, 246; Stanton, 356; Forbes, 223-25, 253.)

†Francis Egerton (1800-57), first Earl of Ellesmere, was a British statesman, art patron, and supporter of scientific societies. Charles Lyell (1797-1875), a British geologist, conducted many scientific expeditions in Europe and is considered the father of modern geology.

Congress reported the expedition and said it had been "successfully carried on."[18] And it most certainly had. Even a cursory glance at its achievements shows how valuable it was. It had, in the first place, made known to the world the exact conditions and the navigability of the Paraná and Paraguay rivers, not to mention their many tributaries. Brazil had already taken advantage of the information and concluded a treaty of navigation to her interior provinces with Paraguay. Because of this discovery too, commerce began to grow rapidly in the Argentine cities of Rosario, Santa Fé, and Paraná particularly. Page concluded that ocean steamers four times the *Water Witch*'s tonnage could ascend these rivers 2000 miles "at *all* seasons of the year."[19] In addition, the Salado River, "never before passed over by the white man," was shown to be navigable for eight hundred miles.[20]

In the area of natural history the contributions of the expedition were enormous. It sent back to the United States the largest collection ever made in South America by an American expedition up to that time. For this, Page said, Dr. Carter deserved most of the credit. The specimens included quadrupeds, birds, fishes, reptiles, insects, crustacea, shells, minerals, plants, seeds, woods and fossils—many of them previously undescribed.[21] John Cassin* at the Academy of Natural Sciences, Philadelphia, where the collection of birds was sent, wrote Page that many of them had been unknown until then. He felt the expedition was "not inferior to the *Beagle*† or d'Orbigny expeditions," and had made contributions to zoological science of national aspect. Charles Girard‡ at the Smithsonian Institution of Washington wrote Page that among the fishes and reptiles there were "several species entirely new to science." Joseph Carson of the University of Pennsylvania admittedly was "impressed with the importance to science of the specimens brought."[22]

As to history, how much could be said! A whole series of "firsts" belongs to the expedition: the first (and only) side-wheel and sail man-of-war, the first combined commercial and scientific expedition to the interior of a foreign country, the first steamer on the Paraná, Paraguay, and other rivers, the first long exploration into the Chaco country, the first steamboat ever constructed and launched on the Paraguay River (the *Pilcomayo*), the first large vessel to go into the Brazilian province of Mato Grosso.

And who can tell what other far-reaching effects the presence of the *Water Witch* had in the La Plata region? What if General Urquiza had been captured or killed in Buenos Aires instead of living to consolidate the Argentine Confederation? What of the vast number of immigrants encouraged by the prospects presented by the opening of the waterways? What of the part played in the history of Bolivia by Ortíz and his companions? In all of these things the role of the *Water Witch* may have been seemingly small, but how great may have been that smallness in reality!

On August 4, 1858, Captain Page sent his final report to Dobbin. He concluded with this noble thought:[23]

*Dr. John Cassin (1813-69), a Pennsylvania ornithologist, had examined the specimens of Wilkes' expedition and Perry's expedition to Japan (1852-56). He also looked at Herndon's materials but no money was appropriated to work on them. (Ponko, 30, 92, 158.)

†A British scientific expedition sailed on the *Beagle*, 1831-36, to the South Pacific, Australasia, and the coasts of South America. It provided Charles Darwin (1809-82), its naturalist, much of the evidence he needed for his theory of evolution. For d'Orbigny see n., p. 21.

‡Dr. Charles F. Girard (1822-95), a French ichthyologist, came to America in 1847 with Louis Agassiz. He had also examined the specimens of Wilkes' expedition as well as the U.S. Army expedition to the Great Salt Lake (1852) and the Zuni and Colorado Rivers (1854). Dr. Joseph Carson (1808-76) as a physician specialized in teaching the medicinal value of natural products.

> Sufficient has been accomplished . . . to induce the sanguine hope, that through the humble means of this Expedition, sent abroad solely under the authority of the Department, there may result to this and those countries an important commerce, and to science, valuable contributions in both Geography and Natural History.*

*A modern historian has judged Page to be wholly deserving of these expressed hopes. "Page planned well, but was not able always to utilize his equipment in the most advantageous way," said Ponko, perhaps referring to nagging mechanical difficulties with his ships and to constrictions imposed by South American governments. But he agreed that Page had effected "a better climate for commerce . . . [and had made] the La Plata and its tributaries known to the people who lived there as well as to the United States . . . a pioneering effort . . . a lasting achievement." Page's two expeditions, Ponko concluded, "accomplished a feat of reporting on the geographical features and economic resources of the region superior to that of any other nineteenth-century expedition to South America." (Ponko, 109, 132-33.)

Use of the lasso and bolas.

12. DRIFTAGE IN THE WAKE

No story ever really ends. It always has aspects that continue into time, echoes and reverberations which grow faint with the years, perhaps, but are still heard. And if the story is absorbing or important, we wonder at the end of it what became of the principal characters and how the unsolved or unresolved matters were resolved. In this sense there were many things left, so to speak, in the wake of the *Water Witch*. There were, first of all, three things the rupture with Paraguay had left unfinished: the unratified treaty, reimbursements to the American company, and indemnification to the family of Chaney, along with some sort of apology for the Itapirú incident.

The day after Page's final report was completed Secretary of State Marcy started negotiations with Richard Fitzpatrick, who had been appointed special agent to Paraguay. He told him that the ratification exchange deadline had been extended by the Senate to "any time within the discretion of the President."[1] Fitzpatrick relayed this information to the Paraguayan Minister of Foreign Affairs, Nicolás Vásquez, who replied on November 8 that his country would have to know first where the United States stood in regard to the Hopkins and *Water Witch* affairs before the exchange could be made. Furthermore, there was a minor problem. The treaty of 1853 had been amended by the United States Senate which, so far as Paraguay was concerned, invalidated it; another would have to be drafted in its place.[2]

The treaty of 1853 throughout had contained the words "United States of North America." The word "North" had to be dropped as not properly belonging to the official title of the country.[3] It was inconsequential in a way, but quibbling went on all through 1857 until President Buchanan, in his first annual message to Congress in December asked for authority to deal with Paraguay by force, if necessary.[4] In February 1858 Bamberger, still consul in Asunción, wrote to the new Secretary of State, Lewis Cass, that through Brazil Paraguay had opened the Paraguay River "to the Commerce of the world," and that it was no longer necessary to push the treaty matter.[5] But Cass wanted a treaty—and he also wanted a settlement of the Hopkins affair. On June 2 Congress passed a joint resolution:[6]

> . . . for the purpose of adjusting the differences between the United States and the republic of Paraguay in connection with the attack upon the U.S. Steamer Water Witch . . . the President is hereby authorized to adopt such measures and use such force as in his judgment may be necessary and advisable in the event of a refusal of just satisfaction by the Government of Paraguay.

Then, as if to force Paraguay's position, on October 5 the *Water Witch* was sent on a "good will mission" to Asunción.[7] To back up this gesture James B. Bowlin was commissioned by Cass to get an indemnity for the American company. He was told he could settle for half a million dollars outright; if Paraguay declined he was to appoint an impartial commission to settle the matter, then get a treaty. The document was to be ratified in Paraguay first and sent to Washington for its ratification within twelve months.[8] As a still further show of American "interest" a naval expedition of nineteen armed vessels, carrying two hundred guns and 2500 men was sent to the La Plata.[9]

President Urquiza of Argentina was evidently alarmed by the arrival of an impressive naval force.* On January 4, 1859, Bowlin wrote from Rosario saying there was a rumor that Urquiza was going to Asunción, presumably to favor the United States mission. It proved to be true, and Bowlin actually met Urquiza in Paraná on the tenth. The President reminded him he owed his life to the United States and wanted to help. He said he felt he could bring about the necessary negotiations himself without the help of Brazil or Montevideo, through which the American diplomats had thought of mediating.[10]

On January 24 Bowlin arrived in Asunción and went to the Minister of Foreign Affairs to arrange a meeting with President López. To his surprise and pleasure he discovered that Urquiza was living right next door to him. The Argentine President invited him over and indicated that things looked promising.[11] Two days later Bowlin, Commodore Shubrick,† and nine officers were cordially received by López and Urquiza, and the Brazilian minister assured Bowlin that everything would soon be settled.[12] Meanwhile, in Paraná, General Santa Cruz called on the United States Minister in Argentina, Benjamin C. Yancey, to say he hoped Bowlin would demand free navigation of the Paraguay, Pilcomayo, and Bermejo rivers, and that he was sure Bolivia and Argentina would back up the demand.[13] But he was too late.

On February 4 a treaty based on the one of 1853 was signed.[14] At the same time a special convention was authorized to provide for the appointment of a commission to settle the Hopkins

*Urquiza was unsure of the fleet's real purpose, one that might jeopardize his plans regarding the unification of Argentina, still his major problem. Washburn dismissed Urquiza with South America's other many tyrannical *caudillos* (leader, commander): "This man, who from poverty and obscurity, by a life of fraud, rapacity, and robbery, had obtained possession of the better, if not the larger, part of the province of Entre Rios, so that he always had a large army under his control, had been at one time the ally and supporter of Rosas in his infamous and blood-thirsty career." Urquiza nevertheless seems the only sane and rational figure in the stormy politics of the La Plata. His equanimity undoubtedly commended itself to Page, whose trust and friendship were long and enduring. Urquiza was assassinated in 1870. (Washburn, i 381.)

†William B. Shubrick (1790-1874) had commanded the Pacific squadron in the Mexican War. He was also a skillful diplomat in this affair—and long remembered for it, although the $3 million expedition—the largest U.S. naval force up to that time—failed its mission (see n., p. 85). "And it is equally certain," Washburn charged, with Southern treachery fresh in his mind, "that the sailing of this expedition was made a pretext for withdrawing from the forts and arsenals of the North all the munitions of war, thus leaving them unprovided with arms whenever the plans for the Great Rebellion should be matured. . . ." (Washburn, i 378; Peterson, 258.)

affair,* to agree to the payment of $10,000 to the family of Samuel Chaney, to give permission for a United States exploring vessel to enter each of the rivers in Paraguay, and to offer a full letter of explanation and apology for the assault on the *Water Witch.*†[15] The signing of the treaty and the convention was celebrated by the singing of a *Te Deum* in the cathedral and bands playing in the streets.[16] The vessel *Tacuarí* took Urquiza back to Argentina on February 5.[17]

*Hopkins did not participate in any of the 1858-59 negotiations. From about 1855 he lived the remainder of his life in and around Buenos Aires, where he vigorously continued his steamship, railroad, telegraph, and other civic works, always with some quasi-official diplomatic appointment in the Argentine government. (Hopkins had always maintained—probably with some acuity—that in Latin America one needed official status to succeed commercially.) Curiously, he was never far from Page (n., p. 86), whose post-Civil War activities in Argentina must have been similar. Hopkins revealed that President López in 1861 had tried to have him assassinated; his claims against Paraguay were never resolved.

In 1888 Hopkins was in New York addressing the Chamber of Commerce—as he had the Geographical Society 36 years before—on "The Argentine Republic and Coterminous Countries," and at the same time constructing dredges in Baltimore for deepening the waterways between Argentina and Paraguay. He died in 1891.

(Washburn; Ynsfran, 249-55; Harrison, 194; Peterson, 258-61.)

†López shrewdly insisted the Chaney indemnity and his official apology for the ship's attack be kept secret, so he would not be embarrassed at home. The Hopkins commission, instead of awarding a settlement, reopened discussion on the claim's legitimacy; even President Buchanan was disgusted about the whole matter. The affair was strange indeed.

Commissioner Bowlin had engaged a New Yorker, Samuel Ward (1814-84), as secretary and translator for his mission to Paraguay. A friend of many politicians and other notables, Ward had been brilliant in mathematics, spoke several languages, and was attracting attention in Washington as a "lobbyist" of the first magnitude. But in the autumn of 1857 Ward for all his precocity was not yet a rich man. Paraguay would change that.

As Bowlin performed his task Ward privately met President López and accepted a secret lobbying assignment. He was to use his influence (1) to secure ratification of the commercial treaty and convention; and (2) to reduce the $500,000 indemnity to Hopkins' company to the lowest figure possible. For this last he was to receive 2 percent of the net savings to López. They agreed upon Spanish pseudonyms by which they would correspond. On February 13, 1859, Bowlin and his official party departed from Asunción.

In Buenos Aires Ward learned that Hopkins, noisily condemning Bowlin's settlement, had intimated the existence of a plot to assassinate López. (Possibly Ynsfran confused cat and mouse.) In Montevideo he found that Shubrick and his naval staff had generally concurred unofficially that the half-million dollar indemnity sounded much too high. These cogent items he communicated to López—at the same time requesting payments on account.

Ward recommended a British lawyer (actually suggested by Shubrick) to represent Paraguay in the indemnity proceedings, and contrived to have himself named a voting member in case of deadlock. In New York he ingratiated himself with a representative of the Rhode Island company, and at the same time (May 1859) persuaded the *New York Times* to discontinue a series of Hopkins' letters attempting to discredit Bowlin. The company agent attempted to employ him on its side, but Ward demurred.

Finally, in March 1860 the treaty and convention were ratified. Ward quickly pocketed a final payment (total 1000 pounds sterling) for this part of his assignment. When the indemnity commission met (June 1860) Ward was secretary and interpreter. After two months of argumentation even the U.S. government commissioner (a personal friend of President Buchanan) favored the defendant, and on August 13 the decision was announced: ". . . the said [Paraguayan] Government is not responsible. . . ." The Rhode Island company was to get nothing.

Buchanan was so annoyed—having sent the expensive fleet he now looked foolish—he claimed the commission had exceeded its jurisdiction. President Lincoln next sent Washburn (see n., p. 63) to Paraguay to see if the matter could be reopened, but the Civil War soon swept the affair away. (Still later attempts to recover damages were also fruitless.)

Pablo Max Ynsfran (1894-1972), Paraguayan historian and statesman (exiled in 1940 to the United States, where he taught at the University of Texas), reconstructed these details, mostly from the pseudonymous Spanish letters, which he identified in the archives. Ynsfran did not doubt that Sam Ward got the $10,000 "honorarium" for this part of the bargain—in 1860 he was apparently spending money lavishly on exquisite dinner parties—but he also believed Ward extremely lucky, his influence minimal on two events that would have been resolved, much as they were, without his aid. The Paraguayan case was stronger than participants at first thought possible.

Upon its conclusion Bowlin wrote to López: ". . . You claimed to demand nothing but equal and exact Justice; and I had abiding faith and confidence in my Countrymen. . . ." López, delighted with his sly triumph, did not mind at all if Page returned to continue his scientific exploration of Paraguayan waters.

(Pablo Max Ynsfran, "Sam Ward's Bargain with President López of Paraguay." *Hispanic American Historical Review*, Vol. 34, August 1954, 313-331.)

On March 10 the seventy-two year old López held a reception for twenty-nine officers, most of them from the *Water Witch.** Among them was Murdaugh, whom everyone was happy to see. "Mr. Murdaugh they regarded as a brother and all seemed eager to grasp him by his hands." López wore a cross of diamonds for the occasion.[18] It was rather symbolic, all things considered. When the *Water Witch* left there was peace for all men of good will.†

And what of the men who explored and worked and guided the *Water Witch* through three years of expeditionary effort? What became of them?

Second Assistant Engineer William Lamdin was promoted through First Assistant to Chief Engineer on October 1, 1861. He had a long naval career, retiring in 1887. He died on October 12, 1888. Second Assistant Engineers T.B.C. Stump and R. C. Potts were both promoted to First Assistants on May 9, 1857. Apparently their sympathies in the Civil War lay with the South, for both were "stricken from the role" in 1861, the former on May 22 and the latter on June 15.

Among the officers too there were divided loyalties. When the war broke out Ammen talked to Lieutenant William Powell, who was then stationed under Maury at the Naval Observatory.‡ Powell told him he was against officers leaving the Navy because Abraham Lincoln had become president.[19] His sentiments, however, must have changed, because he was dismissed from the Navy on April 20, 1861. The next day Lieutenant William Murdaugh was dismissed and the day after that Thomas J. Page, who had been raised to the rank of commander while still on the expedition (September 14, 1855), and who was in fact and in heart a Virginian, was also dismissed from the service.§

*Despite its separate "good will mission" the *Water Witch* was now actually part of Shubrick's fleet. Page was not its commander; he was captain of the *Salinas*, the fleet's flagship. Murdaugh (below) was apparently still assigned to the *Water Witch*.

†After the fleet returned home Page proceeded to the La Plata area to commence his second expedition. The *Argentina*'s boiler had been faulty, however, and her draft deeper than calculated. Forbes replaced the boiler and built another—smaller—ship, the *Alpha*; Page at Buenos Aires accepted the two vessels in March, 1859. (Forbes received $300 a month for them; when he later sold them he reported he had lost $8000 on the whole transaction.)

On October 6, 1860, Page in Buenos Aires announced his field work finished. The two ships were returned to Forbes' representative and he and his officers booked commercial passage to New York. By January 1861 Page was at work putting his records and charts in order—but the results of the second expedition were never published. When Fort Sumter was bombarded on April 12, 1861, Page's research—and his career in the United States Navy—were effectively ended.

(Ponko, 131; Forbes.)

‡Neither Ammen nor Page revealed that Maury, whose brilliant work must have impressed scores of young naval officers —themselves included, had two almost unbelievable escape schemes for his fellow Southerners.

As early as 1849 sensing the enormity of North-South differences over slavery, Maury supposed that Southern slave-holders could either colonize the Amazon Valley or sell their slaves to plantation owners already there, as Brazil had just opened the Amazon to (domestic) steam navigation. They would thereby avoid great financial loss through emancipation. It was, said Harrison, "probably the most extreme version of Manifest Destiny ever suggested by a public official."

Maury was energetic in using Herndon's findings (n., p. 9) to publicize the agricultural potential of the Amazon and other South American valleys. (Was he collecting soil samples with this in mind?) He had also surmised that strong ocean currents between Brazil and the Atlantic states would be excellent for brisk trade. But Brazil did not open the river to foreign vessels until 1867 (and abolished slavery in 1888); by then, of course, steam navigation had shown itself able to overcome the whims of wind and current. Maury, a loyal Virginian, had also resigned his commission in 1861, his naval career over.

Maury went to Mexico at the invitation of Emperor Maximilian to become Imperial Commissioner of Emigration. He attempted especially to establish a colony of Virginian émigrés; they introduced the cultivation of cinchona for the cure of malaria, but the Emperor quickly lost interest in the scheme. Maury in 1866 sailed to England, where he was welcomed for his scientific work, but two years later he returned to America to teach meteorology at Virginia Military Institute.

(Ponko, 61-64, 187-88; Harrison, 197; Maury, "Annual Address," 10-13.)

§Page entered Confederate service as a naval commander and was engaged in the defense of Virginia. In 1863 he was sent as a secret agent to England where he obtained, after months of delay, a powerful French ironclad renamed the *Stonewall*; unfortunately by the time he reached Havana the Civil War had ended, and he turned the ship over to Spanish authorities. Like many proud Confederates he evidently felt it painful or unwise to return home; at first he considered settling in Mexico.

All the others remained in the Navy. Lieutenant James Moore died on October 18, 1860; he was spared the crisis of war. Edmund Henry was made a lieutenant commander on July 16, 1862. He retired from the Navy March 19, 1867 and died May 8, 1872. The three remaining officers all reached high positions and served with distinction. William Jeffers was promoted to lieutenant commander in 1862, commander in 1865, captain in 1870, and commodore in 1878. He died July 23, 1883.* Edward Donaldson, after having been made a commander in 1862, captain in 1866, and commander in 1871, rose to the rank of rear admiral on September 21, 1876. He retired eight days later and died May 15, 1889. Lieutenant Daniel Ammen was made commander and captain the same years as Donaldson. He became a commodore in 1872 and on December 11, 1877, a rear admiral. His book, *The Old Navy and the New*, records his varied and interesting experiences. He died July 11, 1898.†[20]

And the *Water Witch*?

After the La Plata expedition the *Water Witch* was employed for some time in the Caribbean pursuing slave ships.[21] From October 1858 to May 1859 the steamer made her second voyage to Paraguay, previously described. The surgeon who was on this trip did not think much of the ship, calling it "a small box of a steamboat that ought never to have been sent to sea."[22] But her past record had already more than justified her presence on the ocean. When the Civil War erupted the *Water Witch* was under the command of a Kentuckian, Lieutenant Commander Austin Pendergast, with seventy-seven in her crew.[23] Her work was mostly on the blockade of Southern ports. She had at least one battle—with the gunboat *Ivy*—which resulted in a draw.

On June 2, 1863, Pendergast anchored the steamer in Ossabaw Sound, off the coast of Georgia, a half mile north of Bradley's River. The ship was fairly close to shore, and that night a thick fog blanketed the area.[24] Under its cover a hundred and fifty Confederates silently filled seven cotton barges and stealthily approached the little man-of-war.[25] Low-sided vessel that she was, the *Water Witch* was easily boarded and in a matter of minutes a melee ensued. The Confederates were soon in control. Of the ship's crew two were dead, fourteen wounded, and sixty-one captured. Among the attackers were six dead, including the commander, and seventeen wounded. Among the men captured on board only one got away[26]—one who, like Ishmael in *Moby Dick*, could quote the Book of Job: "And I only am escaped alone to tell thee."

What did the Confederates do with the *Water Witch*? The naval records give no account. She seems to have mysteriously disappeared. Perhaps, true to her name, she brewed her own particular magic—and vanished into history.

Instead, Page returned to Argentina, where he spent some time on the cattle ranch in Entre Rios province belonging to his old friend, ex-president Urquiza. Page had noticed, when his son John attended the Concepción college founded by Urquiza in 1850, that when its income was inadequate Urquiza himself paid its expenses; Page admired that. In 1859 Page had requested permission to present Urquiza a field howitzer from the *Argentina*, but this was denied.

The Argentine government sent Page to England to superintend the construction of two ironclads and two gunboats for a new fleet—and a new escalation of Latin American wars. In 1880 he moved his family to Florence, Italy. He died 1899 in Rome.

(Ponko, 111, 131, 133; *Dictionary of American Biography*, xiii, 140-41.)

*Jeffers commanded the famous ironclad *Monitor* for a time during the Civil War. After two years at the Washington Naval Observatory he was promoted to captain. He then became Chief of the U.S. Navy Bureau of Ordnance, 1873-83, where he designed and produced bronze and steel boat howitzers and made other technical improvements to guns and related materiél. (*National Cyclopedia of American Biography*, iv 281.)

†Ammen (1819-98) remained at the Naval Observatory (see n., p. 40) for two years. In the Civil War he participated in the naval attacks on Port Royal (1861), Fort McAllister (1863), and Fort Fisher (1864-65). A Northerner—a neighbor and good friend of President Grant, Ammen was of course unaffected by Maury's views. In 1869 he became Chief of the Bureau of Navigation and sent a number of surveying expeditions to Central America, and was himself involved in early discussions on canals in Nicaragua and Panama. He nevertheless expressed no further personal interest in Latin America. (Ponko, 179.)

REFERENCES

CHAPTER 1: OVERTURE TO AN ADVENTURE

1. Burton Stevenson, Ed., *The Home Book of Quotations* (New York: Dodd, Mead, 1967).
2. National Archives, General Records of the Department of State (RG 59), Consular despatches, 1789-1906.
3. Ricardo Levene, *A History of Argentina*, William S. Robertson, Trans., Ed. (Chapel Hill: U. North Carolina Press, 1937), 416-28; Harris G. Warren, *Paraguay: An Informal History* (Norman: U. Oklahoma Press, 1949), 185; Joao Pandía Calógeras, *A History of Brazil*, P. A. Martin, Trans., Ed. (New York: Russell & Russell), 164-82.
4. Warren, 190.
5. Pablo Max Ynsfran, *La Expedición norteamericana contra el Paraguay, 1858-1859* (Buenos Aires: Guarania, 1954), 42.
6. Warren, 190.
7. Charles A. Washburn, *A History of Paraguay*, 2 Vols. (Boston: Lee & Shepard, 1871), i 354.
8. Warren, 190.
9. William Manning, Ed., *Diplomatic Correspondence of the United States* (Washington: Carnegie Endowment for International Peace, 1932), iii 29-32, document 4500.
10. Ynsfran, 45; Manning, 55, document 4516.
11. Ynsfran, 45.
12. Ibid., 46.
13. Manning, 59-61, Hopkins to Buchanan, Sept. 16, 1845.
14. Ibid., 56, Hopkins to Buchanan, Aug. 5, 1845.
15. Ynsfran, 46.
16. Manning, 63, Hopkins to Buchanan, Nov. 31 [sic], 1845; Ynsfran, 41.
17. Manning, 65, Hopkins to Buchanan, Nov. 31, 1845.
18. Ibid., 66.
19. Ibid., 80, Hopkins to Buchanan, Feb. (no date) 1846.
20. Ibid., 68, Hopkins to Buchanan, Nov. 31, 1845.
21. Ibid., 68, Hopkins to Buchanan, Feb. 12, 1846.
22. Ibid., 67, Hopkins to Buchanan, Nov. 31, 1845.
23. Ibid., 82, Hopkins to Buchanan, Feb. (no date) 1846.
24. Ibid., 67, Hopkins to Buchanan, Nov. 31, 1845.
25. Warren, 191.
26. Manning, 80, Hopkins to Buchanan, Feb. (no date) 1846.
27. Ibid., 75, Hopkins to Buchanan, Nov. 31, 1845.
28. Ibid., 77, Hopkins to Buchanan, Feb. 12, 1846.
29. Ynsfran, 78.
30. Ibid., 79.
31. Manning, 88-89, Hopkins to López, March 27, 1846.
32. Ibid.
33. Ibid., 89-90, Buchanan to Hopkins, March 30, 1846.
34. Ibid., 90-91, Hopkins to López, April 20, 1846.
35. Ibid., 91-95, Hopkins to Buchanan, June 10, 1846.
36. Warren, 191-92.
37. Levene, 437-39.
38. Levene, 443; Warren, 193.
39. Manning, 34, Webster to Pendleton, April 28, 1852.
40. Ynsfran, 162.
41. Thomas Jefferson Page, *La Plata, the Argentine Confederation, and Paraguay* (New York: Harper, 1859), xxii. Hereafter cited simply as Page.
42. Warren, 192.
43. Daniel Ammen, *The Old Navy and the New* (Philadelphia: Lippincott, 1891), 244. Hereafter cited simply as Ammen.
44. Page, xix; Ammen, 245.
45. Washburn, i 364; Ynsfran, 162-63.
46. National Archives, Military and Naval Branch, Letters to Officers of Ships of War No. 47, April 20, 1852-Oct. 27, 1852, 418-19.
47. Page, 567.
48. Ibid., 567-68; Ynsfran, 161.
49. Ammen, 245.
50. Frank M. Bennett, *The Steam Navy of the United States* (Pittsburgh: Warren, 1896), 51-52; Ynsfran, 162.
51. Ynsfran, 162; Page, xx.
52. Page, xxi.
53. National Archives, Military and Naval Branch, Letters to Officers of Ships of War No. 48, Oct. 27, 1852-May 30, 1853, 74.
54. Ibid., 122, 124.
55. Ibid., 91-92.
56. Ibid., 199-200.
57. Ibid., 34.
58. Page, 302.
59. Ibid., 302.
60. National Archives, Military and Naval Branch, *Letters &c. from Lieutenant Thomas J. Page*, Jan. 6, 1853 and Aug. 4, 1856 (Record Group 45, Navy Department, Letters from Officers Commanding Expeditions), under Jan. 23. Hereafter cited as Page, "Letters."
61. Ibid., under Feb. 5, 1853.
62. Ibid., under Jan. 20, 1853.
63. Page, "Letters," under Jan. 21; Letters to Officers of Ships of War No. 48, 234.
64. Letters to Officers of Ships of War No. 48, 247, 248.
65. Manning, 35, Everett to Page, Feb. 1, 1853.
66. Page, xxii.
67. "Log Book No. 5, Water Witch," Jan. 12, 1853-Dec. 29, 1854 (National Archives, Military and Naval Branch, Record Group 24).

CHAPTER 2: SIDE-WHEELS SOUTHWARD

1. "Log Book No. 5," entries Feb. 17-23, 1853.
2. Page, "Letters," Feb. 18, 1853.
3. Ammen, 245.
4. "Log Book No. 5," entries Feb. 24-March 4, 1853; Page, "Letters," March 10, 1853.
5. Ammen, 245-46.
6. Page, "Letters," March 10, 1853.
7. "Log Book No. 5," entries March 13, 16, 17, 1853.
8. Ammen, 246-47.
9. Letters Received Naval Observatory, file box 23

(National Archives, Military and Naval Branch, Record Group 78), Page to Maury, April 25, 1853.

10. "Log Book No. 5," entries March 24-April 7, 1853.
11. Ammen, 249.
12. Ibid., 249; "Log Book No. 5," April 14, 15, 1853.
13. Page, "Letters," April 16, 1853.
14. "Log Book No. 5," under April 17, 1853.
15. Ibid., April 22, 23, 1853.
16. Letters Received Naval Observatory, file box 23, Page, April 25, 1853.
17. Ibid.
18. Thomas Jefferson Page, *Report of the Exploration and Survey of the River "La Plata" and Tributaries* (Washington: Wendell, 1856), 4. Hereafter cited as Page, "Report."
19. Manning, 104, C. A. López to President of the U.S., Feb. 26, 1853.
20. Ibid., 107, documents 4535, 4536.
21. Page, "Letters," April 29, 1853.
22. Page, "Report," 5-6.
23. Ammen, 250; Page, 32-33.
24. "Log Book No. 5," April 24-26, 1853.
25. Letters Received Naval Observatory, May 20, 1853, file box 22.
26. Page, "Letters," April 27, 1853.
27. Letters Received Naval Observatory, Page, May 20, 1853, file box 22.
28. Ammen, 250.
29. Page, 30.
30. Letters Received Naval Observatory, Page, May 20, 1853, file box 22.
31. Ammen, 251.
32. "Log Book No. 5," May 8, 13, 21, 1853.
33. Ibid., May 18, 1853.
34. Page, "Report," 6-7.
35. Page, 34; Letters Received Naval Observatory, Page, May 20, 1853, file box 22.
36. Letters Received Naval Observatory, Page, May 30, 1853, file box 22.
37. Ibid.
38. "Log Book No. 5," May 16, 24, 25, 1853.
39. Page, "Letters," May 28, 1853; Levene, 446.
40. Page, 37.
41. Ibid., 39-40.
42. Letters Received Naval Observatory, Page, May 20 (cont. May 28), 1853, file box 22.
43. Page, "Letters," May 28, 1853.
44. Letters Received Naval Observatory, Page, May 20 (cont. May 28), 1853, file box 22.
45. Page, 39-40.
46. "Log Book No. 5," May 26, 31-June 8, 1853.
47. Page, "Letters," June 30, 1853.
48. Page, 43.
49. "Log Book No. 5," June 20, 1853.
50. Letters Received Naval Observatory, Page, May 20 (cont. May 28), 1853, file box 22.
51. Page, "Letters," July 3, 1853; see also Levene, 446-47.
52. Page, "Letters," June 30, July 3, 1853.
53. Ibid., June 24, 1853.
54. Ibid., July 3, 1853; Page, "Reports," 8.
55. Page, 38.
56. "Log Book No. 5," July 4, 1853.
57. Page, "Letters," July 3, 1853.
58. Page, "Letters," Aug. 2, 1853; Page, 47; "Log Book No. 5," July 13, 1853.
59. Page, "Letters," July 10, 1853.
60. Ibid., Aug. 2, 1853; Page, 47; "Log Book No. 5," July 13, 1853.
61. Page, 49.
62. Ibid., 50; "Log Book No. 5," July 15, 1853.
63. "Log Book No. 5," July 18, 1853; Page, "Letters," Aug. 2, 1853.
64. "Log Book No. 5," July 19-21, 1853; Page, 50.
65. Page, "Letters," July 22, 1853; Page, 51.
66. Page, "Letters," Aug. 2, 1853; "Log Book No. 5," July 25, 1853.
67. Ammen, 251-52.
68. Page, 56, 59.
69. Ibid., 61; Page, "Letters," Aug. 2, 1853.
70. Page, "Letters," Aug. 2, 1853.
71. "Log Book No. 5," Aug. 3-9, 1853; Letters to Officers of Ships of War No. 48, 234.
72. "Log Book No. 5," Aug. 8-21, 1853. Direct quote is under Aug. 21.
73. Page, "Letters," Aug. 19, 1853, enclosed in letter to Dobbin dated Aug. 23, 1853.
74. Ibid., Aug. 20, 1853.
75. Ibid., Aug. 21, 1853.
76. Ibid., Aug. 30, 1853.
77. Ammen, 252.
78. "Log Book No. 5," Aug. 25-31, 1853.

CHAPTER 3: UP THE PARANA

1. Ammen, 253.
2. "Log Book No. 5," Sept. 1, 1853.
3. Ammen, "Reference Journal," Aug. 31, 1853-Jan. 30, 1854 (National Archives, Military and Naval Branch, Hydrographic Office Field Notebooks, 1855-1939, Record Group 37, Box 341.6A.18 to 341.6A.37), B341.6A.34, Sept. 2.
4. "Rio Paraná, Note Book 1A," Sept. 1-22, 1853 (National Archives, Military and Naval Branch, Hydrographic Office Field Note Books, 1855-1939, Record Group 37, Box 117), B341.6A.52, Sept. 3. Hereafter cited as "Rio Paraná Note Book."
5. Page, 71.
6. Ammen, 254.
7. Page, 71.
8. "Rio Paraná Note Book," Sept. 5; Ammen, "Reference Journal," Sept. 5, 1853.
9. "Rio Paraná Note Book," Sept. 5, 1853.
10. Page, 70, 67.
11. "Log Book No. 5," Sept. 6, 1853; Ammen, "Reference Journal," Sept. 6, 1853.
12. Ammen, 253.
13. Page, "Reports," 12.
14. Ammen, "Reference Journal," Sept. 7, 1853.
15. Ibid.
16. Page, 78.
17. Ammen, "Reference Journal," Sept. 9, 13, 1853.
18. Ibid., Sept. 14, 1853.
19. Page, 91.
20. Ibid.

21. Ibid., 92; Ammen, "Reference Journal," Sept. 14, 1853.
22. Ammen, "Reference Journal," Sept. 15, 1853.
23. Page, 66-67. The original has "reveld" and "paradisiacal"; I have corrected the spelling rather than add "sic" twice to so short a quotation.
24. Ammen, "Reference Journal," Sept. 16, 1853.
25. Ibid., Sept. 17, 1853.
26. "Rio Paraná Note Book," Sept. 19, 1853.
27. Page, 96.
28. "Log Book No. 5," Sept. 20, 1853.
29. Ammen, "Reference Journal," Sept. 20, 1853.
30. Ibid.
31. Page, 98.
32. Ammen, "Reference Journal," Sept. 20, 1853.
33. "Rio Paraná Note Book," Sept. 22, 1853.
34. Ammen, 255.
35. Ammen, "Reference Journal," Sept. 22, 1853.
36. Ibid.; Page, 103.
37. Ammen, "Reference Journal," Sept. 22.
38. Page, 100.
39. Ammen, "Reference Journal," Sept. 23, 1853.
40. Ibid., Sept. 25, 1853.
41. Page, "Report," 12.
42. "Log Book No. 5," Sept. 26, 1853.
43. Ammen, "Reference Journal," Sept. 26, 1853.
44. Ibid., Sept. 27, 1853; Page, 109.
45. Ammen, 254.
46. Page, "Report," 12.
47. Page, 108; Ammen, 254.
48. Ammen, 254.
49. Ammen, "Reference Journal," Sept. 28, 1853.
50. Ibid., Sept. 29, 1853.
51. Ibid.; Page, 113.
52. Ynsfran, 161.
53. Ammen, "Reference Journal," Oct. 1, 1853; Ammen, 258.
54. Ammen, "Reference Journal," Oct. 9, 1853; Ammen, 258.
55. Page, 118.
56. Ammen, "Reference Journal," Oct. 10, 1853; Page, 119.
57. Ammen, 257; Page, 123.
58. Page, 124.
59. Ibid., 125.
60. Ammen, "Reference Journal," Oct. 9, 1853.
61. "Log Book No. 5," Oct. 11, 1853.
62. "Log Book No. 5," Oct. 12, 1853.
63. Ammen, "Reference Journal," Oct. 9, 15, 1853.
64. "Log Book No. 5," Oct. 17, 1853; Ynsfran, 133-34. Ynsfran has Oct. 11, but the log book, which I consider more accurate and reliable, records the event on the 17th, with the hour, and "18 days from Buenos Aires."
65. Ynsfran, 133.
66. Washburn, i 359-61.
67. "Log Book No. 5," Oct. 9, 10, 17, 18, 30, 1853.
68. Page, "Letters," to Dobbin, Oct. 20, 1853.
69. Ammen, "Reference Journal," Oct. 23, 1853.
70. Ibid., Oct. 30, Nov. 6, 1853.
71. Ammen, 264.
72. "Log Book No. 5," Nov. 4, 1853.
73. Page, 126.
74. Ammen, "Reference Journal," Nov. 6, 1853.

CHAPTER 4: THE HEART OF A CONTINENT

1. Page, 106.
2. Page, "Letters," Oct. 1, 1853.
3. Ibid., Oct. 2, 1853. I have written out "and," for which Page uses "&."
4. Ibid.
5. Ibid., Oct. 20, 1853.
6. Ibid., Oct. 13, 1853.
7. Manning, 111-12, Page to Marcy, Oct. 20, 1853.
8. Page, "Letters," Nov. 5, 1853.
9. Ibid.
10. Page, 128.
11. "Rio Paraná, Note Book No. 2A," Sept. 22, 1853-Nov. 16, 1853 (National Archives, Military and Naval Branch, Hydrographic Office Field Notebooks, 1855-1939, Record Group 37, Box 341.6A.38 to 341.6A.65), 341.6A.53, Nov. 7. The dates indicate that the Notebook is in the wrong box.
12. Letters Received Naval Observatory, Oct. 1, 1854-March 20, 1855, Page, Dec. 1 (no year). While all other Letters Received are in file boxes, this group is in a bound volume in chronological order. The letter, lacking a year, has been bound into this volume by mistake; its contents clearly indicate 1853.
13. Ibid.
14. "Rio Paraná, Note Book No. 2A," Nov. 9, 11, 1853.
15. Page, 133.
16. "Log Book No. 5," Nov. 10, 12, 13, 1853.
17. Page, 139.
18. Ibid., 140.
19. Ibid., 142.
20. "Log Book No. 5," Nov. 15, 1853; Page, "Report," 14.
21. "Rio Paraná, Note Book No. 2A," Nov. 16, 1853; Page, 143.
22. Page, 143-44, 146.
23. Ibid., 145.
24. Ibid., 146.
25. "Log Book No. 5," Nov. 19, 1853.
26. Page, 161.
27. Page, 152-53.
28. Ibid., 162; Log Book No. 5," Nov. 21, 1853.
29. "Log Book No. 5," Nov. 25, 1853; Page, 170, 173.
30. Page, "Report," 17.
31. "Log Book No. 5," Nov. 27, 1853.
32. Page, 180-81; Page, "Report," 15.
33. Letters Received Naval Observatory, bound volume, Page, Dec. 1 (no year).
34. "Log Book No. 5," Nov. 29, 1853.
35. Page, 186.
36. Ibid., 185-86.
37. "Rio Paraná, Note Book No. 3A," Nov. 16, 1853-Nov. 30, 1853 (National Archives, Military and Naval Branch, Hydrographic Office Field Notebooks, 1855-1939, Record Group 37, 341.6A.38 to 341.6A.65), B341.6A.54, Nov. 30, 1853. The dates indicate the Notebook is in the wrong box.
38. Page, "Letters," Dec. 1, 1853; Page, "Report," 14.
39. Page, 187.
40. Page, "Letters," Dec. 1, 1853.
41. Letters Received Naval Observatory, bound volume, Page, Dec. 1 (no year).
42. Ibid.

43. Page, "Report," 14-15; Page, "Letters," Dec. 1, 1853.
44. Page, 189.
45. Page, "Report," 15.
46. Page, 190.
47. Ibid., 191-92.
48. Ibid., 193, 195; "Log Book No. 5," Dec. 7, 1853.
49. Ammen, 259; "Log Book No. 5," Dec. 5, 1853.
50. "Log Book No. 5," Dec. 7, 1853.
51. "Log Book No. 5," Dec. 9, 1853; Page, 194.
52. "Log Book No. 5," Dec. 11, 13, 14, 1853; Page, 196.
53. Page, 196.
54. Ynsfran, 169.
55. Manning, 115, Page to Varela, Dec. 21, 1853.
56. Ibid., 113, Hopkins to Marcy, Dec. 27, 1853.
57. Ibid.
58. Letters Received Naval Observatory, bound volume, Page, Dec. 1 (no year).
59. Page, "Letters," Dec. 23, 1853.
60. Ibid., Jan 26, 1854.
61. "Log Book No. 5," Dec. 23-30, 1853.
62. Page, "Letters," Jan 26, 1854.
63. Ibid.
64. Ammen, "Reference Journal," Nov. 29, Dec. 16, 1853.
65. Page, 246.
66. Ammen, "Reference Journal," Jan. 7, 16, 20; "Log Book No. 5," Jan. 23, 30, 1854.

CHAPTER 5: INLAND AND THREE RIVERS

1. Ammen, "Notebook No. 2," Feb.-May 14, 1854 (National Archives, Military and Naval Branch, Hydrographic Office Field Notebooks, 1855-1939, Record Group 37, Box 116), B341.6A.35, Feb. 6, 1854. The dates indicate this notebook is in the wrong box.
2. Page, 207.
3. Ammen, "Notebook No. 2," Feb. 6, 1854.
4. Ibid., Feb. 7, 1854.
5. Ibid., Feb. 8, 9, 1854.
6. Ibid., Feb. 11, 1854.
7. Page, "Report," 17.
8. Ammen,' "Notebook No. 2," Feb. 13-18, 1854.
9. Ibid., Feb. 22, 1854.
10. Page, "Report," 18.
11. Ammen, "Notebook No. 2," Feb. 23, 25, 1854.
12. William Murdaugh, "Steamer Pilcomayo, Rios Paraguay, Confuso and Salado, Feb. 1854" (National Archives, Military and Naval Branch, Hydrographic Office, Field Notebooks, 1855-1939, Record Group 37, 341.6A.18 to 341.6A.37), B341.6A.33, Feb. 10, 1854. The dates indicate this notebook is in the wrong box.
13. Ibid., Feb. 12, 1854.
14. Ibid., Feb. 13, 14, 1854.
15. Ibid., Feb. 12, 15. 1854.
16. Ibid., Feb. 16-18, 1854.
17. Ibid., Feb. 19, 20, 1854; Ammen, 257.
18. Murdaugh, "Steamer Pilcomayo, Rios Paraguay, Confuso and Salado," Feb. 21-24, 1854.
19. Ammen, "Notebook No. 2," Feb. 26, 1854.
20. Ammen, 263.
21. Ammen, "Notebook No. 2," March 1, 12, 19, 1854.
22. Page, "Letters," March 10, 1854.
23. Ibid.
24. Ibid., Oct. 20, 1853.
25. Page, "Report," 16.
26. Page, "Letters," March 13, 1854.
27. Page, 242.
28. Ammen, "Notebook No. 2," March 26, 1854.
29. Ibid., March 12, 1854.
30. Ibid., April 2, 1854.
31. Ibid., April 11, 1854.
32. Ibid., April 12-25, 1854.
33. Ibid., April 15, 1854.
34. Ibid., April 16, 1854.
35. Ibid., April 17, 1854.
36. Ammen, 261-62.
37. Ammen, "Notebook No. 2," April 18, 1854.
38. Ibid., April 19, 22, 23, 1854.
39. Page, "Letters," May 10, 1854.
40. Ammen, "Notebook No. 2," May 14, 1854.
41. Ammen, 269.
42. Page, 240; Page, "Report," 16.

CHAPTER 6: SUNLIT PLAINS AND PERENNIAL STREAMS

1. "Log Book No. 5," Feb. 9, 10, 1854.
2. Ibid., Feb. 12, 1854.
3. Ibid., spotted entries, Feb. 16-April 1, 1854, especially March 9.
4. Ibid., Feb. 21, 23, 24; March 9, 12, 15, 31; April 1, 15, 1854.
5. Ibid., Feb. 18, and spotted entries, Feb. 24-March 8, 1854.
6. Ibid., Feb. 22, 1854.
7. Ibid., March 14, 25, 1854.
8. Ibid., April 13-May 1, 1854.
9. Ibid., May 11, 1854.
10. Ibid., May 18, 1854.
11. Page, "Letters," July 12, 1854.
12. Page, 247-48, 257.
13. Ibid., 248, 250.
14. Ibid., 248-50.
15. Ibid., 249, 253-54.
16. Page, "Letters," July 12, 1854.
17. Page, 258; Page, "Letters," July 12, 1854.
18. Page, 259.
19. Ibid., 248; Page, "Letters, July 12, 1854.
20. "Log Book No. 5," July 5, 1854.
21. William L. Powell, "Paraguay" (National Archives, Military and Naval Branch, Hydrographic Office Field Notebooks, 1855-1939, Record Group 37, 341.6A.18 to 341.6A.37), 341.6A.32, May 31, June 2, 1854.
22. Ibid., June 8, 11, 12.
23. Ibid., June 13-17, 1854.
24. Powell's "Observations for Positions in the Interior of Paraguay" can be found in the National Archives, Military and Naval Branch, Hydrographic Office Field Notebooks, 1855-1939, Record Group 37, Box 117, marked 341.6A.64. Hereafter cited as Powell, "Paraguay."
25. Powell, "Paraguay," June 20, 26, 1854.
26. Ibid., June 21, 1854.
27. Ibid., June 24, 1854.

28. Ibid., July 1, 1854.
29. Ibid., July 2, 3, 6, 1854.
30. Ibid., July 8, 9, 1854.
31. Ibid., July 11, 12.
32. Ibid., July 23-27, 1854.
33. Ibid., July 29-31, 1854.
34. Page, "Report," 18.
35. Powell, "Paraguay," Aug. 1, 2, 1854; Page, 222.
36. Powell, "Paraguay," Aug. 2, 3, 1854; Page, 222.
37. Page, "Report," 17.
38. Powell, "Paraguay," Aug. 5, 6, 1854.
39. Page, 226.
40. Powell, "Paraguay," Aug. 7, 1854; Page, 227.
41. Page, "Report," 17.
42. Page, 120.
43. Powell, "Paraguay," Aug. 8-15, 1854.
44. Ibid., Aug. 16, 1854; Page, 236.
45. Powell, "Paraguay," Aug. 20, 21, 1854.
46. "Log Book No. 5," July 8, 10, 13, 1854; Page, 264.
47. Page, "Letters," Aug. 13, 1854; "Log Book No. 5," July 14, 1854.
48. Page, "Letters," Aug. 13, 1854; "Log Book No. 5," July 13, 1854.
49. Page, "Letters," July (no date), 1854.
50. "Log Book No. 5," Aug. 3, 1854; Page, "Letters," Aug. 13, 1854; Page, 266.
51. Page, "Letters," Aug. 13, 1854; Page, 265-66.
52. "Log Book No. 5," Aug. 13, 1854; Page, "Letters," Aug. 13, 1854.
53. Page, "Letters," Aug. 13, 1854.
54. Page, 264.
55. Ibid.; Page, "Report," 22.
56. Letters Received Naval Observatory, Page, Oct. 2, 1854; Page, "Report," 22.
57. Page, 268.
58. Page, "Letters," Sept. 1, 1854.
59. "Log Book No. 5," Sept. 9, 18, 20, 1854.

CHAPTER 7: TROUBLE AT ASUNCION

1. Manning, 126, Hopkins to Marcy, Aug. 22, 1854; Ynsfran, 175-77.
2. Manning, 118, Falcón to Hopkins, Aug. 5, 1854.
3. Ynsfran, 189.
4. Manning, 119, Hopkins to Marcy, Aug. 12, 1854.
5. Ynsfran, 189-90.
6. Manning, 120, Hopkins to Falcón, Aug. 16, 1854.
7. Manning, 121-33, Hopkins to Marcy, Aug. 22, 1854.
8. Manning, 133, Hopkins to Marcy, Aug. 25, 1854; Washburn, i 367.
9. Manning, 137, Hopkins to Marcy, Sept. 1, 1854.
10. Manning, 138-39, Falcón to Marcy, Sept. 2, 1854.
11. Letters Received Naval Observatory, Page, Sept. 26, 1854, file box 28.
12. Ynsfran, 191.
13. Page, "Letters," Sept. 29, 1854; Ynsfran, 194-95.
14. Letters Received Naval Observatory, Page, Sept. 26, 1854, file box 28.
15. Page, "Letters," Sept. 28, 1854 (two letters); Ynsfran, 196-97.
16. Page, "Letters," Sept. 29, 1854; Ynsfran, 197-98.
17. Page, "Letters," Sept. 19, 1854.
18. Ibid.
19. "Log Book No. 5," Sept. 28, 1854.
20. Ibid., Sept. 29, 1854.
21. Page, "Letters," Oct. 2, 1854; Washburn, i 369.
22. Letters Received Naval Observatory, Page, Oct. 2 (no year), bound volume.
23. "Log Book No. 5," Sept. 30, 1854; Page, 275.
24. "Log Book No. 5," Sept. 30, 1854.
25. Ibid., Oct. 2, 1854; Letters Received Naval Observatory, Page, Oct. 2 (no year), bound volume.
26. Page, "Letters," Sept. 29, 1854, Oct. 2, 1854.
27. "Log Book No. 5," Oct. 3, 1854.
28. Ibid., Oct. 5, 1854; Page, "Letters," Sept. 29, 1854.
29. Manning, 141-42, Falcón to Marcy, Oct. 3, 1854.
30. Letters Received Naval Observatory, Page, Sept. 26, 1854, file box 28.
31. Ibid., Page, Oct. 26, 1854, bound volume.
32. Page, "Letters," Sept. 29, 1854.
33. Letters Received Naval Observatory, Page, Oct. 26, 1854, bound volume.
34. Ibid.
35. Page, "Letters," Oct. 17, 1854; Manning, 142-43, Page to Falcón, Oct. 16, 1854.
36. "Log Book No. 5," Oct. 16, 1854.
37. Manning, 144, Falcón to Page, Oct. 21, 1854; Page, "Letters," Nov. 5, 1854.
38. Manning, 145-57, Page to Marcy, Nov. 5, 1854.
39. Page, "Letters," Nov. 5, 1854.
40. Manning, 148-49, Hopkins to Marcy, Dec. 9, 1854.

CHAPTER 8: GUNFIRE AT ITAPIRU

1. "Log Book No. 5," Oct. 9, 18, 19; Nov. 12, 1854.
2. Ibid., Nov. 13, 15, 17, 19, 1854.
3. Ibid., Nov. 20, 21, 28; Dec. 6, 16, 21, 1854.
4. Ibid., Dec. 23-31, 1854.
5. Edward W. Callahan, *List of Officers of the Navy of the United States and the Marine Corps, 1775-1900* (New York: Hamersley, 1901), 294.
6. "Log Book No. 6," Water Witch, Jan. 1, 1855-Dec. 31, 1855 (National Archives, Military and Naval Branch, Record Group 24), Jan. 1, 1855; Ynsfran, 213.
7. Page, 290-91.
8. Ibid., 96, 292-93. Page makes extensive use of Murdaugh's notes throughout this section of his book.
9. Ibid., 295-98.
10. Ibid., 296.
11. Page, "Letters," Jan. 1, 1855.
12. "Log Book No. 6," Jan. 20, 1855; Ynsfran, 213.
13. "Log Book No. 6," Jan. 31, 1855; Page, "Letters," Jan. 30, 1855.
14. Ynsfran, 211.
15. There are three sources for the events of this day: "Log Book No. 6," Feb. 1, 1855; Manning, 154-55, Jeffers to Page, Feb. 2, 1855; Manning, 150, Robles to Carlos A. López, Feb. 1, 1855. I have combined the details of the three sources in order to make a more fluent narrative. The basic account is that of Jeffers.
16. Manning, 151-53, Falcón to Marcy, Feb. 4, 1855.
17. "Log Book No. 6," Feb. 2-5, 1855; Page, 304.

18. Page, "Letters," Feb. 5, 1855.
19. Page, 569.
20. "Log Book No. 6," Feb. 5, 8, 1855.
21. Manning, 154-59, Peden to Marcy, Feb. 10, 1855.
22. Manning, 163, Buenaventura Decoud to Peden, March 21, 1855.
23. Page, "Letters," April 16, 1855; Page, 314-15.
24. Manning, 160, Peden to Marcy, April 6, 1855.
25. Ibid., 161, Peden to Marcy, April 20, 1855.
26. Page, "Letters," April 17, 1855.

CHAPTER 9: THE URUGUAY AND ARGENTINA RIVERS

1. Page, "Letters," Feb. 12, 1855.
2. Ibid.
3. Ibid., Feb. 16, 1855.
4. Ibid., April 8, 1855.
5. "Log Book No. 6," March 5, 6, 9, 16, 20, 24; April 2, 10, 1855.
6. Page, "Letters," March 21, 1855.
7. "Log Book No. 6," April 24, 25, 1855.
8. Page, 320.
9. "Rio Uruguay, Note Book No. 1," Water Witch (National Archives, Military and Naval Branch, Hydrographic Office Field Notebooks, 1855-1939, Record Group 37, Box 118), B341.6A.81, April 27, 1855; "Log Book No. 6," April 27, 1855. There are two Rio Uruguay notebooks, but since Note Book No. 2 will not be cited in this work, Note Book No. 1 will be cited hereafter as "Rio Uruguay."
10. "Rio Uruguay," April 28, 1855; Page, 322.
11. Page, 322; Page, "Report," 22.
12. Page, 323; "Rio Uruguay," May 2, 1855.
13. "Log Book No. 6," May 6, 7, 1855.
14. Page, "Report," 23.
15. Page, 324.
16. "Rio Uruguay," May 7, 8, 1855.
17. Page, "Report," 22.
18. "Log Book No. 6," May 8, 1855; Page, 325.
19. "Rio Uruguay," May 10, 1855.
20. Page, 326; "Rio Uruguay," May 11, 1855.
21. Page, "Report," 23.
22. "Log Book No. 6," May 12, 16, 1855; Page, 321.
23. "Log Book No. 6," May 18-23, 1855.
24. Page, 329; "Log Book No. 6," May 26, 28, 1855.
25. Page, "Report," 23; Page, 330-331.
26. Page, "Letters," June 19, 1855.
27. Letters Received Naval Observatory, Page, June 24, 1855, file box 29.
28. "Log Book No. 6," June 30, July 2, 1855.
29. Page, 322.
30. "Yerba" notebook, Water Witch (National Archives, Military and Naval Branch, Hydrographic Office Field Notebooks, 1855-1939, Record Group 37, 341.3.15 to 341.6A.17), B341.6A.1, July 3, 5, 1855. Hereafter cited as "Yerba."
31. "Yerba," July 7-11, 1855.
32. Page, 72, 333-34.
33. Ibid., 333.
34. Ibid., 335.
35. Ibid., 335; "Rio Salado, Steamer Yerba, Note Book No. 1, July 1855," Water Witch (National Archives, Military and Naval Branch, Hydrographic Office Field Notebooks, 1855-1939, Record Group 37, Box 117), B341.6A.65, July 14-16, 1855.
36. Page, 335.
37. "Rio Salado, Note Book No. 1," July 17, 18, 1855.
38. Page, 335.
39. Ibid., 336; "Rio Salado, Note Book No. 1," July 18, 1855.
40. Page, 336; "Rio Salado, Note Book No. 1," July 20, 21, 1855.
41. Page, 337-38.
42. "Rio Salado, Steamer Yerba, Note Book No. 2, July & August 1855," Water Witch (National Archives, Military and Naval Branch, Hydrographic Office Field Notebooks, 1855-1939, Record Group 37, Box 118), B341.6A.66, July 25, 1855.
43. "Rio Salado, Note Book No. 2," July 26-28, 1855.
44. Ibid., July 29, 1855; Page, 340.
45. "Rio Salado, Note Book No. 2," Aug. 6-11, 1855.
46. "A list of the birds skinned in the Rio Salado during the month of July, 1855," in "Rio Salado, Santa Fe," etc. (National Archives, Military and Naval Branch, Hydrographic Office Field Notebooks, 1855-1939, Record Group 37, Box 118), B341.6A.77.
47. Page, 343.
48. "Riacho Paranacito" (National Archives, Military and Naval Branch, Hydrographic Office Notebooks, 1855-1939, Record Group 37, Box 117), 341.6A.61, Aug. 15, 20, 23, 25, 27, 29, 1855; "Log Book No. 6," Aug. 29, 1855.

CHAPTER 10: MOMENTOUS DISCOVERIES

1. Page, "Letters," Aug. 7, 1855.
2. Page, 343-44; Page, "Report," 26.
3. Page, 344, 348; Page, "Report," 26.
4. Page, 346-47, 349-51.
5. Ibid., 353-55, 358-59.
6. Ibid., 361-63.
7. Ibid., 362, 365, 368, 370, 372; Page, "Report," 26.
8. Page, 375, 377-78, 387-91.
9. Page, "Letters," Oct. 22, 1855.
10. Page, 398-403.
11. Ibid., 407-408, 411, 413-14.
12. Ibid., 416.
13. Ibid., 417-18, 420, 425, 429; Page, "Report," 27.
14. Page, "Report," 30. Bushell is probably the translator.
15. Ibid.
16. Page, "Letters," Dec. 15, 1855.
17. "Log Book No. 6," Aug. 3, 1855; Page, "Report," 10.
18. "Log Book No. 6," Aug. 22, 1855.
19. "Log Book No. 6," Oct. 26-29; Page, "Letters," Dec. 15, 1855.
20. "Log Book No. 6," Oct. 31, 1855.
21. Page, "Report," 10.
22. Page, "Letters," Dec. 15, 1855.
23. Letters Received Naval Observatory, Page, Dec. 16, 1855, file box 31.
24. "Log Book No. 6," Nov. 6, 8, 10, 12; Dec. 5, 1855.
25. Manning, 163, Bamberger to Marcy, Dec. (no date), 1855.

CHAPTER 11: THE END OF THE EXPEDITION

1. Page, "Letters," Dec. 19, 1855.
2. "Log Book No. 7," Water Witch, Jan. 1, 1856-May 12, 1856 (National Archives, Military and Naval Branch, Record Group 24), Jan. 14, 16, 19, 23, 1856.
3. Page, "Report," 23.
4. Page, "Letters," Jan. 28, 1856.
5. "Log Book No. 7," Feb. 3, 6, 7, 10, 12, 1856.
6. Page, "Letters," Feb. 15, 1856.
7. "Log Book No. 7," Feb. 12, 15, 18, 20, 25.
8. Page, "Letters," April 1, 1856.
9. "Log Book No. 7," March 2, 3, 11, 1856.
10. Ibid., March 11, 17, 19, 22, 1856.
11. Ibid., March 22, 29, 1856; Page, "Letters," April 8, Feb. 23, 1856.
12. Page, "Letters," May 7, 1856.
13. "Log Book No. 7," April 10, 16, 19, 21, 24, 1856.
14. Ibid., April 26, 29, 30, 1856.
15. Ibid., May 4-7, 1856; Page, "Letters," May 7, 1856.
16. Ibid., May 8-12, 1856.
17. Page, "Report," 33.
18. James Buchanan, "First Annual Message." *Messages and Papers of the Presidents* (New York: Bureau of National Literature, 1897), vii 2980.
19. Page, "Report," 13, 33.
20. Ibid., 33.
21. Ibid., 31.
22. Ibid., 36-38.
23. Page, "Letters," Aug. 4, 1856.

CHAPTER 12: DRIFTAGE IN THE WAKE

1. Manning, 37-38, Marcy to Fitzpatrick, Aug. 5, 11, 1856.
2. Manning, 171-72, 174, Vásquez to Fitzpatrick, Nov. 8, 15, 1856.
3. Washburn, i 371-72.
4. *Messages and Papers of the Presidents*, vii 2980.
5. Manning, 183, Bamberger to Cass, Feb. 15, 1858.
6. John Mabry Mathews, *American Foreign Relations* (New York: Century, 1928), 544-45.
7. Charles T. Fahs, "Diary." Library of Congress, Manuscript Division (Vol. ac. 6431), 6.
8. Manning, 37-46, Cass to Bowlin, Oct. 6, 1858.
9. Buchanan, "Third Annual Message." *Messages and Papers of the Presidents*, vii 3091.
10. Manning, 192-200, Bowlin to Cass, Jan. 4, 10, 1859.
11. Ibid., 200-201, Bowlin to Cass, Jan. 25, 1859.
12. Ibid., 204, Yancey to Cass, Feb. 4, 1859.
13. Ibid.
14. Warren, 194.
15. Manning, 208-11, Bowlin to Cass, Feb. 17, 1859.
16. Fahs, 102.
17. Manning, 204-205, Yancey to Cass, Feb. 5, 1859.
18. Fahs, 99, 101.
19. Ammen, 332.
20. Callahan is official source for biographical data: Lamdin, 320; Stump, 528; Potts, 443; Powell, 443; Murdaugh, 398; Page, 418; Moore, 388; Henry, 261; Jeffers, 294; Donaldson, 166; Ammen, 23.
21. Ynsfran, 171.
22. Fahs, 6.
23. Virgil C. Jones, *The Civil War at Sea* (New York: Holt, Rinehart, Winston, 1962), iii 206.
24. Ibid.
25. Bennett, 369.
26. For these and other details of the capture, see Jones, Chap. 15, "The Bewitched Witch," 206-214.

ESSAY ON SOURCES

Only one printed account of the *Water Witch* expedition in any way comes close to being a detailed narrative. This is Thomas Jefferson Page's *La Plata, the Argentine Confederation, and Paraguay* (New York: Harper & Brothers, 1859). Even then, it fails to record many details, as the present work shows, and about 200 of its 632 pages are merely a history of the La Plata region. It is, however, the only source of information for the second half of the Salado River exploration (Chapter 10). A summary of the expedition was written by Page shortly after his return and was published under the title *Report of the Exploration and Survey of the River "La Plata" and Tributaries by Thomas J. Page, Commanding the United States Steamer Water Witch, to the Secretary of the Navy, 1856* (Washington: Wendell, 1856). In several cases the facts recorded in Page's two works are in disagreement; in such instances I have used those of the "Report," since it was written immediately upon Page's return and is probably more accurate. Page undoubtedly relied on his memory a great deal for his longer work, and in three years details can become hazy.

A third contemporary source for the trip is the partial chapter in Rear Admiral Daniel Ammen's *The Old Navy and the New* (Philadelphia: Lippincott, 1891). This work proved valuable for details of occurrences in places where Page merely mentions the names, particularly on the outbound voyage. The expedition is also referred to in general in Pablo Max Ynsfran's *La Expedición norteamericana contra el Paraguay, 1858-1859*, 2 Vols. (Buenos Aires: Guarania, 1954-58). As the title suggests, this work deals mainly with the retaliatory expedition described in Chapter 12. Volume I was, however, a good source on Hopkins which I used for Chapters 1 and 7. (Ynsfran's second volume was not used.) The same general delineation of the trip is contained in Charles A. Washburn's *The History of Paraguay, with Notes of Personal Observations, and Reminiscences of Diplomacy under Difficulties*, 2 Vols. (Boston: Lee & Shepard, 1871), in two chapters of Volume I. Vincent Ponko, Jr.'s *Ships, Seas, and Scientists: U.S. Naval Exploration and Discovery in the Nineteenth Century* (Annapolis: Naval Institute Press, 1974) devotes a chapter to the trip, "Page Expedition to the Rio de la Plata and Rio Paraguay, 1853-56," from a professional Navy point of view.

Most of what might be termed the "inside information" on the expedition was obtained from documents in the Military and Naval Branch Office of the National Archives, Washington, D.C. Record Group 45, "Letters from Officers Commanding Expeditions," contains a bound volume entitled *Letters &c. from Lt. T. J. Page, commanding U.S. Steamer Water Witch: Exploration and Survey of the Rivers La Plata, Paraguay and Parana and their Tributaries. January 6, 1853 and August 4, 1856.* It was perhaps the prime source for innumerable interesting details and revealed much of Page's character. These letters were written to Secretaries of the Navy John Kennedy and James Dobbin. In this same record group is what might be called a companion section, "Letters to Officers Commanding Expeditions." Under this heading are several volumes entitled "Letters to Officers of Ships of War," containing the scrivener's copies of the letters sent. In No. 47 (April 30-Oct. 27, 1852) and No. 48 (Oct. 27, 1852-May 30, 1853) I found several letters to Page.

Record Group 78 contains "Records of the Naval Observatory." In this section are "Letters Received, Naval Observatory." These, with the exception of one bound volume (Oct. 1, 1854-March 20, 1855), are kept in metal file boxes or drawers according to periods of several months, but are not in any chronological order. One must thumb through thousands of letters to find what is wanted. In this section several of Page's letters to Maury, then in charge of the Naval Observatory, were found. Like the letters to Dobbin, they are valuable for their on-the-spot personal impressions and recordings of events.

Under "Records of the Bureau of Naval Personnel," Record Group 24, are the three log books of the *Water Witch* during this expedition. From them were obtained information on the weather, some incidental events, and times and places of arrival and departure. The present work could hardly have been written without them. They are: No. 5, Jan. 12, 1853-Dec. 31, 1854; No. 6, Jan. 1-Dec. 31, 1855; and No. 7, Jan. 1-May 12, 1856.

The greatest variety of information as to what occurred during the expedition was found in Record Group 37, "Records of the Hydrographic Office." In this section is a series of cardboard boxes labeled "Hydrographic Office Field Notebooks, 1855-1939."

A collection of notebooks, journals, etc., from the second expedition of the *Water Witch* may be found here, and all of those from the first expedition have been mistakenly added to this collection. In box No. 115 are two notebooks entitled simply "Water Witch" and "Yerba," both of which were used. From Box No. 116 information was obtained from the notebooks entitled "Reference Journal" of Lt. Ammen, and his "Notebook No. 2," "Paraguay," by Lt. Powell, and "Steamer Pilcomayo: Rios Paraguay, Confuso and Salado, Feb. 1854," by Lt. Murdaugh. Box No. 117 contains "Rio Parana Note Book 1A" as well as 2A and 3A. These cover the periods respectively of Sept. 1-22, 1853, Sept. 22-Nov. 16, 1853, and Nov. 16-30, 1853. In this box also are "Riacho Paranacito" and "Rio Salado: Steamer Yerba, Note Book 1." Box No. 118 has Note Book No. 2 of the Rio Salado trip, another notebook entitled "Rio Salado, Santa Fe, etc.," and "Rio Uruguay, Note Book No. 1." All these constituted little pieces in putting together the over-all picture of the expedition. The only other journal used was the diary of Charles T. Fahs, kept from Sept. 2, 1858 to May 1859. Fahs was the *Water Witch* surgeon during her "good will" expedition. A few facts from the diary were used in Chapter 12. The diary is kept in the Manuscript Division of the Library of Congress (ac. 6431).

For general historical background I relied on three main sources: for Argentina, Ricardo Levene's *A History of Argentina*, translated and edited by William Spence Robertson (Chapel Hill: University of North Carolina Press, 1937); for Paraguay, Harris Gaylord Warren's *Paraguay: An Informal History* (Norman: University of Oklahoma Press, 1949); and for Brazil, Joao Pandía Calógeras' *A History of Brazil*, translated and edited by Percy Alvin Martin (New York: Russell & Russell, 1963). Two additional reference works on Paraguay are of value: *Historical Dictionary of Paraguay*, by Charles J. Kolinski (Metuchen: Scarecrow Press, 1973), and *Area Handbook for Paraguay* by Thomas E. Weil, Jan Knippers Black, Howard I. Blutstein, David S. McMorris, Frederick P. Munson, and Charles Townsend (Washington: Department of the Army, 1972).

Although the turbulent history of Paraguay could not be presented here in detail, many recurring elements from prehispanic and colonial periods have made it necessary to explain something of their origins. Among many works on the Jesuits' use of reducciones, Philip Caraman's *The Lost Paradise: An Account of the Jesuits in Paraguay, 1607-1768* (London: Sidgwick & Jackson, 1975) is excellent. For Spanish-colonial government, Samuel H. Wilcocke's *History of the Viceroyalty of Buenos Ayres . . .* (London: Symonds, 1807) is useful, as is Thomas J. Hutchinson's *The Paraná: With Incidents of the Paraguayan War, and South American Recollections, from 1861 to 1868* (London: Stanford, 1868). For the nineteenth century see Alyn Brodsky's *Madame Lynch & Friend: The True Account of an Irish Adventuress and the Dictator of Paraguay who Destroyed that American Nation* (New York: Harper & Row, 1975), and a curious account of Paraguay's first industrialization attempts by a prominent Spanish author and historian, Josefina Plá, *The British in Paraguay, 1850-1870*, translated and introduced by Brian C. MacDermot (Surrey and Oxford: Richmond Publishing Company and St. Anthony's College, 1976).

Coverage of European and United States exploration in South America is somewhat scattered and uneven. For a start see Victor Wolfgang von Hagen's *South America Called Them [La Condamine, Humboldt, Darwin, Spruce]* (New York: Knopf, 1943). For accounts of some American expeditions see William Stanton's *The Great United States Exploring Expedition of 1838-1842* (Berkeley: University of California Press, 1975) and William Lewis Herndon's *Exploration of the Valley of the Amazon*, edited by Hamilton Basso (New York: McGraw-Hill, 1952); the latter is a short popularization of Herndon's original 1853-54 two-volume report. A rather long journal article is illuminating: John P. Harrison, "Science and Politics: Origins and Objectives of Mid-Nineteenth Century Government Expeditions to Latin America" (*Hispanic American Historical Review*, Vol. 35, 1955, 175-202). A more recent visit to the La Plata area is W. J. Holland's *To the River Plate and Back: The Narrative of a Scientific Mission to South America, with Observations upon Things Seen and Suggested* (New York: Putnam, 1913); it shows very little had changed during Paraguay's first one hundred years of independence.

Indispensable for an understanding of personalities and objectives in the events described here are Edward A. Hopkins' "My Life-Record, written at the Request of my Nephew, Fred'k Camp, for the Information of Our Family" (Appendix II in Ynsfran, 249-55), and Harold F. Peterson's "Edward A.

Hopkins: A Pioneer Promoter in Paraguay" (*Hispanic American Historical Review*, Vol. 22, 1942, 245-61). Very instructive for the period are Edward A. Hopkins' "Memoir on the Geography, History, Productions, and Trade of Paraguay" (American Geographical and Statistical Society, *Bulletin*, Vol. I, Jan. 1852, 14-72); S. deWitt Bloodgood's "Memorial to the Secretary of the Navy" (American Geographical and Statistical Society, *Bulletin*, Vol. I, May, 1852, 66-72); and Matthew F. Maury's "Annual Address" (American Geographical and Statistical Society, *Bulletin*, Vol. I [sic] Feb. 1854, 3-31).

Interesting and valuable information was obtained from diplomatic sources, particularly William Manning's edited *Diplomatic Correspondence of the United States* (Washington: Carnegie Endowment for International Peace, 1932), Vol. III. Chapters 1, 7, and 12 rely heavily on this source. Some little additional data were obtained from John Mabry Mathews' *American Foreign Relations* (New York: Century, 1928) with regard to the 1858-59 treaty negotiations (Chapter 12), as well as from the *Messages and Papers of the Presidents* (New York: Bureau of National Literature, 1897) on the same subject, using Volumes VI and VII.

Lastly, in regard to the *Water Witch* and her crew, three printed sources were of some assistance. Frank Bennett's *The Steam Navy of the United States* (Pittsburgh: Warren, 1896) contained information on the construction and capture of the *Water Witch*. A more detailed account of the capture may be found in Chapter 15 of *The Civil War at Sea* (New York: Holt, Rinehart, Winston, 1962), Volume III, by Virgil Carrington Jones. For the origin and disposition of the *Argentina* and *Alpha* see Robert B. Forbes' *Personal Reminiscences* (Boston: 1876). The short biographical details of the officers used in Chapter 12 were from Edward W. Callahan's *List of Officers of the Navy of the United States and the Marine Corps, 1775-1900* (New York: Hamersly, 1901).

INDEX